Apollo Mission Control
The Making of a National Historic Landmark

More information about this series at http://www.springer.com/series/4097

Other Springer-Praxis books by Manfred "Dutch" von Ehrenfried

Stratonauts: Pioneers Venturing into the Stratosphere, 2014
ISBN:978-3-319-02900-9

The Birth of NASA: The Work of the Space Task Group,
America's First True Space Pioneers, 2016
ISBN:978-3-319-28426-2

Exploring the Martian Moons: A Human Mission to Deimos and Phobos, 2017
ISBN:978-3-319-52699-7

Manfred "Dutch" von Ehrenfried

Apollo Mission Control

The Making of a National Historic Landmark

 Springer

Published in association with
Praxis Publishing
Chichester, UK

Manfred "Dutch" von Ehrenfried
Leander, Texas, USA

SPRINGER-PRAXIS BOOKS IN SPACE EXPLORATION

Springer Praxis Books
ISBN 978-3-319-76683-6   ISBN 978-3-319-76684-3   (eBook)
https://doi.org/10.1007/978-3-319-76684-3

Library of Congress Control Number: 2018939417

Cover design by Jim Wilkie. Image credit: NASA
Project Editor: David M. Harland

Printed on acid-free paper

This Springer imprint is published by the registered company Springer International Publishing AG part of Springer Nature.
The registered company address is: Gewerbestrasse 11, 6330 Cham, Switzerland

Contents

The Apollo Flight Directors

Chris Kraft   John Hodge   Gene Kranz   Glynn Lunney

Gerry Griffin   Cliff Charlesworth   Milt Windler   Pete Frank

Phil Shaffer   Don Puddy   Neil Hutchinson   Chuck Lewis

Dedication

By the summer of 1962, most of us from the Space Task Group at Langley had moved to Houston. All of us in Flight Operations monitored the construction of Building 30 at the Manned Spacecraft Center with great anticipation, for in the building were two new Mission Control Centers. We eagerly looked forward to flying Gemini out of there in 1965. Little did we know that two decades later, in 1985, one of the control rooms where we worked would be declared a National Historic Landmark. The National Park Service defined the landmark as the third floor Mission Operations Control Room (called MOCR 2) along with four of the surrounding support rooms. They called it the "Apollo Mission Control Center" because most of the Apollo flights, including all the lunar landings, were flown out of the third floor control room.

This book is dedicated to all those who worked in the Mission Control Center; not only the flight controllers but all those who supported the many missions that were flown out of that building up to and including 1992. That includes Gemini, Apollo, Skylab, Apollo-Soyuz, and the early Space Shuttle flights. But the Mission Operations Control Rooms where merely the tip of the iceberg, so to speak. Building 30 was filled with computers and equipment of all kinds and required hundreds of people to make it all come together in order to fly just one spacecraft into Earth orbit, let alone to fly multiple spacecraft to the Moon. It has been conservatively estimated that over its operational lifetime from 1964 to 1992, over five thousand people were involved in making it all happen in this one building. Other people at the Manned Spacecraft Center in other buildings were also working to support these missions.

It is the intent of this book to remember those who supported these spaceflights from the "Apollo Mission Control Center" – as many as we can by name either in the body of the book or in an appendix of Mission Manning Lists. Sadly, many are long gone and those that remain are in their seventies, eighties, and even nineties. This book is published as part of the celebration to commemorate the first Apollo Lunar Landing of July 20, 1969.

Acknowledgments

There are those who championed the cause of saving the Apollo Mission Control Center from the ravages of time and the obsolescence of technology. There were others who would have gutted the building to assign it to more mundane purposes. But there is something about a place that controlled many of the flights of the Apollo Program, including all the lunar landings, that urges its preservation. Those who first made the effort, and deserve formal acknowledgment, were the "Historical Preservationists." We must thank the management of the National Park Service for hiring Dr. Harry Butowsky in 1981 to study the situation. His report in 1984 nominated the control center as a possible National Historic Landmark. This was approved in 1985. This immediately triggered the protection of the Secretary of the Interior, who imposed Standards for the Treatment of Historic Properties. No longer could NASA reassign the historic place for other purposes. So our thanks go to the people in the Department of Interior and the National Park Service who championed this cause.

With the formal designation, other preservation organizations got involved. In Chapter 9 on the restoration of the Apollo Mission Control Center, many of those people are identified. Many in the long list (some of whom have supplied input to this book) deserve further recognition here. They include:

Johnson Space Center
    Eugene F. Kranz, Flight Director
    Edward Fendell, Flight Controller
    Sandra J. Tetley, JSC Historic Preservation Officer
    Jim Thornton, Restoration Working Group Project Manager
    Jennifer Ross-Nazzal, JSC History Office

Space Center Houston
    William Harris, CEO
    Tracy Lamm, COO
    Kim Parker, VP of Development

City of Webster Gift and Challenge
Donna Rogers, Mayor
Wayne Sabo, City Manager
Michael Rodgers, Treasurer
The Hoteliers of Webster, Texas

Kickstarter (American Public Benefit Corporation)
For backing the Webster Challenge

Manned Spaceflight Operations Association (MSOA)
Board of Directors for their support to the restoration and those who served
as well as support for this book

Historical Preservation Organizations
Maren Bzdek, Colorado State University
Wayne Donaldson, Advisory Council on Historic Preservation

Website Support
Jeff Hanley and Jason Roberts, The Aerospace Corporation

The following have provided input to the book, listed alphabetically by subject area:

Auxiliary Computer Room (ACR)/Mission Planning and Analysis Division
Hal Beck
William Sullivan
Elric McHenry
Jerry Bell
Larry Davis
Ken Young

Apollo Lunar Surface Experiments Package (ALSEP) and Lunar Orbit
Burt Sharp
Bill Brizzolara

Command, Communications and Telemetry Support (CCATS)
Tom Sheehan
Ed Pickett

Cosmosphere Spaceworks Restoration Contractor
Jim Remar
Jack Graber
Shannon Whetzel
Dale Capps
Don Aich

National Archives and Records Administration Fort Worth, Texas
Rodney Krajca
Ketina Taylor

Photos
   John Aaron
   Jerry Bostick
   Chuck Deiterich and front cover explanation
   Ed Fendell
   Mary Ann Harness
   Frank van Rensselaer

Poetry
   Spencer Gardner

Real Time Computer Complex (RTCC)
   Neil Hutchinson

Recovery Operations Control Room (ROCR) and People
   John Stonesifer
   Richard Snyder
   George Richeson
   Milt Heflin

Restoration Input and Significance
   Gene Kranz
   Gerry Griffin
   Jeff Hanley
   Bill Reeves

Spacecraft Analysis (SPAN) and Mission Evaluation Room (MER)
   Arnie Aldrich
   Gary Johnson

Staff Support Rooms (SSRs)
   Jerry Bostick, Flight Dynamics SSR
   Larry Keyser, Flight Directors SSR
   Charles Harlan, Flight Directors SSR

University of Houston at Clear Lake Archives
   Lauren Myers
   Jean Grant

Women Flight Controllers
   Linda Ham
   Marianne Dyson
   Jeff Hanley
   Milt Heflin

Many thanks to Wikipedia and Google, which I accessed frequently and was pleased to find that it enabled me to fill in the pieces of the puzzle for just about any subject. Their input is woven into many sections. I was able to find references to the Mission Control Center that are now over half a century old but very useful.

And many thanks to my Springer colleagues Maury Solomon and Hannah Kaufman in New York, Clive Horwood of Praxis in Chichester, England, and cover designer Jim Wilkie in Guildford, England.

I offer a special thanks to David M. Harland in Glasgow, Scotland, who has edited all of my Springer-Praxis books over the past five years. While we have communicated by email, we have never met. Nevertheless, I think of him as a trusted friend. In a way I also think of him as a diamond polisher who, bending over a grind stone, brings out the light and sparkle from an otherwise ugly rock.

Preface

Thank God there are people who want to protect historical places, as without them where would people go to learn about the past. Without them, would we not, over time, just forget? In some cases, the pieces of history can be gathered up and put in museums; a plane here, a rocket there. In every case, those pieces alone do not tell the whole story. A piece of rock from the Moon glimpsed through a barrier of glass cannot begin to tell even a sentence of the story. In other cases, there may be a desire to save the whole, only to find that no one can afford to save it or to keep up with its maintenance. Certainly, preserving a piece of history as significant as mankind's initial step into the void and unknowns of space is one of those places.

There is such a place, well protected from the elements, and in a rather unique government building that is not the whole story of man's quest for the knowledge of space, but at least represents where it all happened. It's a place which very few people have seen with their own eyes, although it will be familiar on their TV sets and in photographs. Wouldn't it be grand to see it in its original light; to see it as it was half a century ago when teams of very talented people controlled the flights that took astronauts to our nearest neighbor, the Moon. Tourists will one day sit in the Visitor Viewing Room and look through the glass wall out onto to the rows of consoles and at the "big screen" on the wall opposite, and see the actual room just as it was during the lunar landings. "See that Johnny, that's where they controlled flights to the Moon when your grandmother and I where your age."

That place is called the "Apollo Mission Control Center." Due to the efforts of many dedicated people it is now a National Historic Landmark, a designation that affords it some Federal protection. But that alone will not restore it to its original configuration and condition. However, it is a start. The government cannot spend money to restore it, but can support the efforts of others who can raise the money. It costs millions to restore a complicated facility like this to its original condition, especially after decades of neglect. But that is the story which I want to tell here, with the help of those who have served in it over many space programs.

In parallel with telling the story of the restoration, I wanted to explain things about the control center that aren't well known to the general public. In fact even though I worked in the control center, in carrying out the necessary research and communicating with colleagues who worked there, I learned some things also! It is unfortunate that the pictures which most people have seen show just those in the Mission Operations Control Room. There were many people in other rooms and other buildings that supported the flight controllers. I have managed to find pictures of people in these other rooms that most people have never seen; even space aficionados.

I also wanted to point out the people who were pioneers in spaceflight; those who developed the operational concepts, methods, and procedures for conducting spaceflight. For the most part they were there in the beginning. And they often came up with the ideas by just "brain storming" on the basis of what little spaceflight experience they had gained from Project Mercury. Such people supported all functional and operational areas, and there is a chapter about them. Some are well-known but you probably have never heard of others. It has been conservatively estimated that over the operational lifetime of the Apollo Mission Control Center, over five thousand people supported all of the missions that were flown from the beginning of the Gemini Program until 1992, when the control center was abandoned. There is a chapter that describes how that came to be.

The work to restore this iconic place is now underway. The famous consoles are being removed from the control room and sent for restoration. The floors and ceilings will be restored and half a century of dirt and smoke cleaned off. Only a few other adjacent rooms that will be visible to visitors will be restored. Overall, the restoration will take a little over a year. The plan is to have the work done in time to celebrate the 50th anniversary of the first lunar landing on July 20, 2019.

The story will start with a brief history of how mission control got started and how the specifications for the new Houston Mission Control Center were drawn up. Each area will be described in the context of the missions, and will describe what the support rooms did to support those in the control center that were seen on TV. Visitors will see what it took to fly a mission.

There are many appendices with relevant material such as key correspondence, mission manning lists, women flight controllers, and other interesting topics and stories with photographs.

This book highlights the people who supported NASA space missions, and is for them and their prodigy as much as it is for students of spaceflight who might one day be inspired to become the flight controllers, engineers, and scientists of the future. Enjoy the story of the Making of a National Historic Landmark called Apollo Mission Control.

Leander, TX, USA   Manfred "Dutch" von Ehrenfried
June, 2018

1

Introduction

In the 1300s, Geoffrey Chaucer said "Time and Tide Wait for No Man." While we can't stop the ravages of time, we can certainly stop and try to capture what was "a truly great" and important period of time in America's history. In this case, not only a great period in America's history, but of a noble activity called human spaceflight. And that unique place is called the Mission Control Center.

Nowadays known as the Johnson Space Center (JSC) the Manned Spacecraft Center (MSC) was opened in 1964. Its location led to it also being referred to as the Houston Mission Control Center (MCC-H). It served its function by giving operational support for the Gemini, Apollo, Apollo-Soyuz, Skylab, and the early Space Shuttle flights. It is the work to preserve some of these facilities that is the subject of this book.

Building 30 is a five story tall structure with three floors that consists of the Mission Operations Wing, the Operations Support (later redesignated the Administrative Wing) and an interconnecting Lobby Wing. It is the Mission Operations Wing from which space missions were controlled. It is here that the world sees NASA conducting spaceflight out in the open, and it is here, and especially the third floor, that has become of historical importance.

Three basic functional areas known as the Display/Control System, the Real Time Computing Complex (RTCC), and the Communications, Command and Telemetry System (CCATS) supported the primary task of conducting manned spaceflight. In the public's eye, this is the Mission Control Center.

The first floor housed the RTCC, the CCATS, and other equipment areas. It supported operations in the two upper floors and also interfaced to the Manned Space Flight Network and the Cape Canaveral launch facilities and down range tracking stations. The Display/Control System, which is distributed throughout the building, is the input/output of the other two systems.

The two floors above housed the Mission Operation Control Rooms (known then as MOCRs, and pronounced moh-kers). The one on the second floor was MOCR 1 and the one on the third floor was MOCR 2. There were many other rooms that supported the

© Springer International Publishing AG, part of Springer Nature 2018
M. von Ehrenfried, *Apollo Mission Control*, Springer Praxis Books,
https://doi.org/10.1007/978-3-319-76684-3_1

MOCRs, including all the functional areas providing data to the flight controllers in the MOCRs. The MOCR was quite simply "the place" for making spaceflight history.

This book will focus primarily on the third floor MOCR 2 because it is this area and surrounding rooms that, following two years of study by the National Park Service (NPS), led on December 24, 1985 to the Department of Interior notifying NASA Administrator James M. Beggs that the Apollo Mission Control Center was to be designated as a National Historic Landmark (NHL).

From MOCR 2 on the third floor came the black-and-white grainy images on 600 million people's television sets during the first lunar landing on the evening of July 20, 1969 and all of the landings that followed. In addition to most Apollo flights, recognition also extends to most of the Gemini and Space Shuttle flights up through STS-53 in 1992. Many unmanned Apollo flights were controlled from MOCR 1, as were Apollo 7, four Skylab missions, and many Shuttle flights. But, as far as the restoration of the NHL is concerned, the period of significance runs from 1964 to the end of the Apollo lunar missions in 1972.

MOCR 2 has not supported missions since 1992, over a quarter of a century ago. Time, as well as new and different missions and advanced technology have made the original contents of the building, including the MOCRs, obsolete. The original facilities and equipment no longer support current missions. NASA has retired them "in situ," not with a "pension" but with a "penchant" for preserving at least MOCR 2 in recognition of its contribution to America's space program. Its new purpose will be to enable visitors to inspect up close and personal what was previously only available to television viewers two generations ago during the "Golden Age of Space." It will serve to demonstrate to a new generation of youngsters from all nations what young people can accomplish; for many of us were only in our twenties at that time. And hopefully this historic landmark will also demonstrate to those in Congress what happened long ago when Presidents (of both parties) and the Congresses made things happen in the national interest.

This book briefly takes us back to the origins of the American space program, to describe how the concepts of spaceflight and mission operations came about, and why and how the Houston Mission Control Center was built. It describes the rooms and positions for both operations and support personnel, and the consoles and displays that they operated. It goes behind the scenes that were shown on TV and describes the facilities and equipment that supplied the flight operations team with the data and functionality they required to support those amazing (and now iconic) missions. It is the place where we heard those words and phrases that are now part of our language and psyche, such as "Houston, Tranquility Base here, the Eagle has landed" and "Houston, we've had a problem." When astronauts in space called home, they were talking to flight controllers in what the NPS calls the "Apollo Mission Control Center." This is what many of us are eager to save from, as Chaucer so elegantly put it, the ravages of time. Those same people are now calling out to others for help. Thousands of people all across the world are donating funds to help with the restoration of these historic facilities. Yet more money will most likely be needed.

The book describes how, during more than half a century since MCC-H was built as a state-of-the-art facility, space missions have changed and why it could no longer support future activities without extensive modifications in hardware, software and operational concepts. It was quite simply time to move on! This led to the eventual abandonment of the third floor MOCR 2. After the lunar missions ended in 1972, it was deactivated and modified for the Space Shuttle, reopening in 1982. It continued to support missions until 1992, when it was abandoned for good. The facility was spent; it was time for others to carry on the nation's space program. The room had seen it all; from the momentous to the tragic.

The rooms lay empty for years and were then used for storage, and for visitor tours and social gatherings. This led to some console degradation and even theft of items. In some places, rips in the carpets were hastily covered over with duct tape. JSC provided power and air conditioning to the facility, but undertook no maintenance and took little interest in its deteriorating condition.

As former flight director Gene Kranz observed, "The overall condition is not emblematic of a National Historic Landmark."

While some people at JSC focused on newer control facilities, now known as Flight Control Rooms (FCR), others were concerned about the historical value of Building 30 and especially its MOCRs.

This book describes the efforts to restore to a presentable level those portions of MOCR 2 and some of the surrounding support rooms and facilities. It focuses on the research, the contracting, the raising of funds, and the industrial team that was contracted to undertake the work. By 2014, a formal working group of only 14 people gathered to plan the proposed work. This included people from various JSC organizations such as the facility and property people, the History Office, the Planning & Integration Office, former NASA flight controllers, as well as people from the National Park Service and the Public Lands History Center at Colorado State University and subcontractors.

By June 2015, an Interagency Agreement initiated by NASA and the NPS had prepared various documents to initiate the necessary work. One of the reports was the Historic Furnishings Report and the Visitor Experience Plan. This defined the depth of detail needed during the restoration of the rooms, consoles and displays, furnishings, and even personal effects of former flight controllers. Five rooms are included: MOCR 2, the Display Projection Room, the Simulation Control Room, the Recovery Control Room, and the Visitor Viewing Area. The report described how the restoration would affect the visitors' experiences – plural because formal tours involving many different groups are envisaged in the Visitor Viewing Room overlooking MOCR 2. Ideally, all this work will be completed well in advance of the 50th Anniversary in July 2019 of the Lunar Landing. The plan is that stories will be told during the tours by experienced people and historic videos played in the restored National Historic Landmark. In this way, memories of the "Golden Age of Spaceflight" will live on.

Since 2017, a JSC Apollo MCC NHL Restoration Project Working Group has routinely met to monitor and manage the restoration effort. Also that year, former NASA flight controllers and astronauts were interviewed; many of them seated at their former consoles in order to provide the contractors with vital information to be used in the restoration process. In order to make the environment as realistic as possible, people have donated documents, reports, plots, and photos that they had at their consoles. And the archives are being searched for as much information as possible. These include the University of Houston-Clear Lake, Rice University's Woodson Research Center, JSC's History Office and even the National Archives in Fort Worth, Texas. The intent is, as former NASA flight controller

Ed Fendell said, "Apollo Mission Control should be restored to a degree of accuracy that will feel to visitors like the day we walked out."

Also in 2017, Space Center Houston, the official NASA JSC visitor center for human spaceflight activities, launched a $5 million campaign to raise funds for a major restoration of the MOCR. $3.5 million had already been contributed by a generous lead gift from the City of Webster, Texas. Then, during July 20 through August 19, Space Center Houston undertook a crowdfunding project called "The Webster Challenge: Restore Historic Mission Control" on Kickstarter, which is a fund raising platform of the American Public Benefit Corporation. The Webster Challenge invited people from around the world to donate over a 30-day period to raise funds for the restoration. As a result, 4,251 people pledged $506,905 to the Kickstarter. The first $400,000 of this will be generously matched by the City of Webster. Thus, the funding was obtained and the restoration got underway. The Apollo Mission Control Center will be saved from the ravages of time and follies of man. But even that amount of money will not be enough, and efforts continue to raise further funds. If you would like to help, then please contact Space Center Houston. Go to: https://spacecenter.org/support/restore-mission-control/donate-now. We heartily thank all of those who have served!

The book will also include many appendices with details of the history, photos, quotes, names, and work of the people involved in capturing a historic period and place during the Golden Age of Space; the iconic Apollo Mission Control Center.

2

Mission Control Concepts

2.1 FLIGHT OPERATIONS

The term "flight operations" has had a variety of meanings to different groups of people and at different points in time. During and after WW-II it was used by the military and described preparing for, and executing fighter and bomber missions. At the National Advisory Committee on Aeronautics (NACA) Langley Memorial Aeronautical Laboratory during this period it was the testing of both military and civilian aircraft. This involved wind tunnels and the acquisition of large amounts of data. Pilots would fly aircraft in order to acquire data in various portions of the flight envelope. There were not "control centers" as we use the term today; there were people in a room inside the hanger who would gather to monitor the "flight test." And at the Wallops Island Station, the Pilotless Aircraft Research Division launched all manner of unmanned craft in the late 1940s and did not use the term so much as "flight test" because for those flights, once you'd "lit the fuse" it was gone!

The same is true of the NACA Muroc Flight Test Unit (later called the High Speed Flight Test Station) while testing the various X planes in the 1940s. The actual flights were monitored by a small group in a room (probably in a hanger) mostly just looking at telemetry and one or two people communicating with the pilot to ascertain his status and the progress through the flight test plan. Control was in the hands of the pilot; there was no control, so to speak, from the ground.

Even before Sputnik, thoughts about how to control orbital flight was a topic of great interest. While there were plans in 1955 for tracking objects from space as part of the International Geophysical Year (IGY), these were really "tracking stations" not control centers as we think of them today. There was no control of the payload once the missile (and they were missiles) was launched.

After Sputnik, and just after the formation of NASA on October 1, 1958, and the Space Task Group later that month, 34 year old Christopher Columbus Kraft, Jr. was given the responsibility to write a "flight test" plan for an orbital mission. Kraft was from an aircraft flight test world so he was an "operations" guy. In his 2001 book, *Flight: My Life in*

© Springer International Publishing AG, part of Springer Nature 2018
M. von Ehrenfried, *Apollo Mission Control*, Springer Praxis Books,
https://doi.org/10.1007/978-3-319-76684-3_2

Mission Control, he describes his initial thoughts for monitoring and controlling an orbital mission. The word "control" came up often because the situation suddenly included not only the capsule but the Army, Air Force, and Navy, as well as NASA. While one would expect this to be more political than operational, it proved to be both.

The more that Kraft and the Space Task Group (STG) studied the problem of how to control an orbital mission, the more the concepts fell into place. During this same period, the Air Force and Army Corps of Engineers were constructing Receiver Building #3 to operate as a telemetry and data processing facility at the Cape Canaveral Air Force Station in Florida, for the Atlantic Test Range. It was intended that the design of that building would accommodate the new "Mercury Control Center" (MCC). The STG specified the overall design of this building, and in particular its layout. As a result, in the STG Organization charts of 1959 the words "Flight Control" and "Control Central" were used. Some people were being hired for their aircraft flight test and operations experience, and others for their understanding of systems and the manner in which these might be used for analysis of capsule performance and potential problems.

Along the way, ideas were coming from contractors who were designing the world-wide tracking stations, in particular how these could be used to "monitor and control" the mission while the capsule was passing over their site. One such contractor was asked what he thought would be required in the way of a facility for analysis and control of the mission; he thought a desk with three telephones ought to do it! Kraft knew that this was the wrong contractor for his concept of how the Mercury missions would be controlled. He presented those concepts to the Society of Experimental Test Pilots on October 9, 1959. Kraft described the Mercury Control Center concept in terms of very specific positions with duties and responsibilities. These "operational functions" were duly manifested in the consoles and displays that defined the Mercury Control Center facility and the main control room.

This methodology of identifying who has responsibility for what during the various phases of a mission, then specifying the displays and controls that they will require in order to perform their individual duties, still defines spaceflight operations and mission control over half a century later. This is a testament to Kraft's vision.

2.2   MILITARY VERSUS CIVILIAN CONTROL

The management of all the "control" elements of a mission were relatively clear by 1959; or at least workably so. The Army made the Redstone missile and they controlled its launch from the blockhouse. Then NASA controlled the remainder of the flight through to completion. The Air Force owned the Cape Canaveral Air Force Station and controlled its land and facilities as well as the tracking stations down the Atlantic Missile Range (the Eastern Test Range as of May 15, 1964). It included Receiver Building #3 and the Mercury Control Center. There would be a console for an Air Force Range Safety Officer. He would blow up a missile which went astray and endangered populated areas. That was his "control" capability. In addition, Air Force contractors assured the facility was ready for mission support. The Navy controlled the recovery forces – except when Admirals fought with the Air Force Generals over which service's air or sea craft would recover the NASA capsule and its astronaut; as it turned out, that battle was won by the NASA Flight Director. The world-wide tracking stations were owned and controlled by NASA Goddard Space Flight Center, although the land on which they stood was owned by the host nation and they were operating under international agreements and at the good graces of the host. The State Department solved most of those problems. Once Project Mercury got underway, some 20 government agencies had become involved. How do you "control" such a bureaucracy?

Dr. Robert Gilruth, the STG Project Manager, appreciated the importance of the various branches of the military and established liaison positions on this staff for very senior officers, most of them at the rank of Colonel or its equivalent. For the most part they wore civilian clothes, even in the Mercury Control Center. The only men in uniform in the MCC were an Admiral in charge of the USN recovery forces, an Air Force General, and Air Force Captain Henry "Pete" Clements who monitored the network and the Atlantic Missile Range assets. There were also a number of uniformed naval officers in the Recovery room, which was alongside the main control room.

These "turf" questions became less of an operational problem as time went on, and NASA made it abundantly clear to all concerned that in matters involving the safety of the crew, the Flight Director in the "control" center was the one who had the life and death responsibility for the astronaut, and his decision would be final. There was no doubt that NASA ran the Mercury Control Center. Two generations later, the international partners and multitudes of contractors acting in support of the International Space Station greatly complicated the process of making decisions, but the authority of the Flight Director is still very clear to all concerned.

Although the Mercury Control Center was completed in 1958, the STG was not yet ready to occupy it because the organization was in the formative stages. Work went on for the tracking and telemetry capabilities of the building. NASA finally occupied the MCC in late 1960 during preparations for the first Mercury Redstone flight.

This history of operational control has been tested and played out many times in the Mercury Control Center, in the Mission Control Center, and in the Apollo Control Center. This book will discuss that history and the efforts to preserve it for posterity.

3

The Original Mission Control Center

3.1 THE MERCURY CONFIGURATION

As initially configured, the floor plan of the Mercury Control Center essentially matched Kraft's 1959 concept of operations as shown in Figure 3.1. After a few simulations and experience with Mercury Redstone 1 in November 1960, some changes were made.

The Range Safety Observer (Position 6) was no longer needed, as the Range Safety Officer was in direct communications with the Flight Director from the facility on the Cape and had the responsibility to destroy the launch vehicle if it went out of limits. This position was assigned to Gene Kranz as the Operations and Procedures Officer (a role that would later evolve into the Assistant Flight Director). And the Recovery Status Monitor (Position 5) was no longer needed because the entire recovery operation was managed from an adjacent Recovery Operations Room. This position also went to Operations and Procedures. (I first occupied this console for John Glenn's MA-6 flight.) The Recovery Task Force commander (Position 2, usually an Admiral) sat on the upper row, behind these two positions.

The Missile Telemetry Monitor (Position 9) was only used for the unmanned MR-1 test on November 21, 1960. The equivalent Redstone and Atlas monitoring positions were in the launch control areas with direct voice communication to the Flight Director. The Booster Systems Monitor position was later added to monitor the launch vehicles. His seat was next to that of Captain Henry Clements, the Air Force Network monitor who was responsible for the status of the Atlantic Missile Range and the remote sites of the Mercury Space Flight Network.

The Support Control Coordination (Position 11) was for John Hatcher, the RCA Maintenance and Operations person who made sure that all the systems in the building that were required by the flight controllers were functional. These included telemetry, displays, power, communications, lighting, data, voice and teletype. This position was later moved over to the left side of the room, facing the flight controllers. It responded to the Operations and Procedures Officer to assure mission readiness.

© Springer International Publishing AG, part of Springer Nature 2018
M. von Ehrenfried, *Apollo Mission Control*, Springer Praxis Books,
https://doi.org/10.1007/978-3-319-76684-3_3

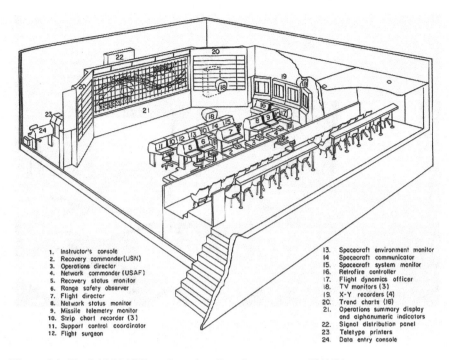

1. Instructor's console
2. Recovery commander (USN)
3. Operations director
4. Network commander (USAF)
5. Recovery status monitor
6. Range safety observer
7. Flight director
8. Network status monitor
9. Missile telemetry monitor
10. Strip chart recorder (3)
11. Support control coordinator
12. Flight surgeon
13. Spacecraft environment monitor
14. Spacecraft communicator
15. Spacecraft system monitor
16. Retrofire controller
17. Flight dynamics officer
18. TV monitors (3)
19. X-Y recorders (4)
20. Trend charts (16)
21. Operations summary display and alphanumeric indicators
22. Signal distribution panel
23. Teletype printers
24. Data entry console

Figure 3.1 The initial MCC configuration. Drawing courtesy of NASA.

Most of the early pictures of the MCC were in black-and-white and somewhat crowded and blurry. Figure 3.2 is a clear, color photo showing how the restored control room looks now that it has been relocated to the Kennedy Center's Early Space Exploration exhibit. The building that housed the MCC was demolished.

3.2 THE GEMINI MODIFICATIONS

After every Mercury mission, ideas for better functional displays were collected for subsequent implementation. Gemini was a two-man spacecraft that flew on a new launch vehicle, the Titan II, so some new positions were added. While the control room functioned about the same as during Mercury, it started to look a little different. For Gemini missions, changes were made to the existing console lineup. Two consoles were added on the left side of the room, facing inward, for the Support Coordinator. And on the right side of the room a fifth plot board was added to the row of four which were used for the Flight Dynamics and Retrofire Officers. More group displays were also added.

Figure 3.2 The Restored Mercury Control Center. Photo courtesy of NASA KSC.

Figure 3.3 Receiver Building #3 housed the MCC. Photo courtesy of NASA.

Figure 3.4 This picture was taken on March 9, 1964 and shows the MCC modified for Gemini. At that time preparations were being made for the launch of Gemini 1, which occurred on April 8. It was also used for Gemini 2, 3, and 4. Photo courtesy of NASA.

This configuration of the MCC continued for the early Gemini missions. The final one to be controlled from this facility was Gemini 3 in March 1965. It was the first manned Gemini with Gus Grissom and John Young. This configuration acted as backup to the new Houston Mission Control Center for Gemini 4, when Ed White made his famous EVA.

Afterwards, the building continued to serve as a remote site tracking station transmitting real time voice, telemetry and data from the Eastern Test Range to the new Houston Mission Control Center. The era of Project Mercury was very definitely over!

3.3 THE TRANSITION TO HOUSTON

Even before John Glenn's flight in February 1962, the plans to move the NASA STG to Houston, Texas were underway. As the STG had just started to support manned Mercury missions, the plans for a new control center were not yet fully developed.

On September 19, 1961 NASA Administrator James E. Webb announced the new Manned Spacecraft Center would be located on the 1000 acre (subsequently 1620 acre) tract south of Houston near Clear Lake. This move from Virginia was not completed until July 1962, but construction work actually began on this tract of cow pastures in April 1962.

Immediately following Glenn's flight, work began on the plans for the new Mission Control Center-Houston. Even the contracts for Gemini and Apollo were not let until the final months of 1962. During this same timeframe, construction was underway at all of the NASA centers and facilities; the new Mission Control Center was just one of many new facilities which were being built around the country at that time.

Even before all of the STG flight operations people relocated to Houston, they began to provide input to the design of the new control center. There were many contractors involved with the buildings, but those most involved with the control center design were picked by a formal procurement process with due inputs from the Operations Division led by Chuck Mathews and Christopher Kraft. Kraft was pleased with the way that IBM and Philco had supported Project Mercury. It was clear that they were the best qualified for the job, and fortunately theirs were the best proposals of the bids by ten companies. After discussions with James Webb and completion of the formal procurement process, IBM and Philco received the contracts. The STG flight control elements most involved with the design effort and with providing input to the winning contractors were the Flight Control and Flight Support Branches of the Operations Division, plus the Mission Planning and Analysis Division. In addition to Kraft, the key players were John D. Hodge, Gerald W. Brewer, Howard C. Kyle, C. Frederick Matthews, Tecwyn Roberts, and John Mayer.

As their duties on Project Mercury wound down, the STG flight controllers started to move to Houston in 1962. There the flight controllers and operations support personnel – the only NASA people with real operations experience of manned spaceflight – focused their efforts on providing inputs to the IBM and Philco contractors. This became a major effort. It resulted in the production of very detailed documents which then drove the console and display designs and support equipment. The results were formally published in the "Flight Control Requirements for the Integrated Mission Control Center."

In the 19 months after the MA-9 flight by Gordon Cooper on May 15, 1963 that wrapped up Project Mercury, Building 30 and the MCC was made ready to conduct mission simulations in support of Gemini.

On December 9, 1964, the partially completed new control center was used passively and in parallel with MCC at the Cape for the Gemini-Titan 2 launch attempt, primarily to validate the computer launch programs. Due to an engine shutdown one second after ignition, the launch had to be scrubbed. The mission finally got underway on January 19, 1965. Considerable utility was made of the telemetry processing program and related TV display formats. The new center received, processed, and displayed live and simulated Gemini launch vehicle and spacecraft data. The results were considered very successful.

The new Houston MCC was backup to the Cape MCC for Gemini 3, which launched on March 23, 1965 when Gus Grissom and John Young flew the first manned Gemini mission. It became the prime control center for the mission of Gemini 4, which launched on June 3, 1965 with Jim McDivitt and Ed White – with the latter making America's first Extra Vehicular Activity (EVA).

It is this same third floor control center that is being honored as a National Historic Landmark and is the subject of this book.

4

A Control Center for the Future

4.1 A NEW CONCEPT

The mental shift from Mercury to Gemini, let alone to Apollo, in the years 1962 and 1963 required some futuristic conceptual thinking. The available technology did not match the sophistication of the spacecraft and the necessary operational concepts. There was something of a disconnect between operations and reality. Operations people are very pragmatic; they contemplate and visualize but don't dream, they rely on scientists and engineers to create the hardware and software they conceive for supporting mission operations. Often, the operational concepts become the requirements that the designer must satisfy. Then too, the operators know that the manufacturers build using their own technologies, therefore flight controllers must learn how systems operate. Operators let others worry about the technology that will go into the spacecraft, launch vehicle and tracking network, and they focus on the control center technology that they will need.

The existing Mercury Space Flight Network of remote sites that provided the control center with data worked fairly well for a one-man spacecraft which flew initially for only three low orbits, mostly over land. This allowed assessment of the astronaut and his "capsule" (as the spacecraft was referred to in those days), and provided "Air/Ground" voice communications when the vehicle was over a remote site station. During a period in contact, information would be passed via voice to the Mercury Control Center, then the "Capsule Communicator" would hastily prepare a message with the pertinent telemetry and doctor's information and send it via low speed teletype (TTY) back to the Mercury Control Center, as illustrated in Figure 4.1. Although this seems rather primitive today, it worked.

Some sites were more capable than others in terms of radars. Only about one third of them had the ability to send a command up to the Mercury capsule. The Bermuda tracking station was a "mini" control center. It had real time capability that could send high speed data via submarine cable to the Goddard Space Flight Center for forwarding to the Mercury Control Center. In fact, it was designed to take operational control of a mission in the event of an accident disabling the Mercury Control Center.

© Springer International Publishing AG, part of Springer Nature 2018
M. von Ehrenfried, *Apollo Mission Control*, Springer Praxis Books,
https://doi.org/10.1007/978-3-319-76684-3_4

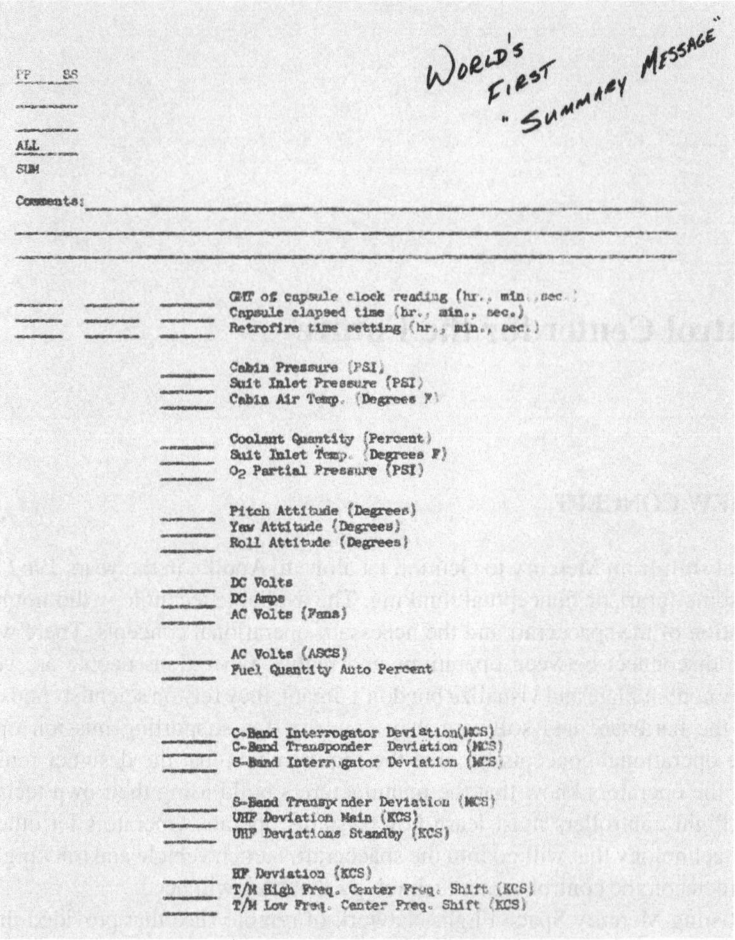

WORLD'S FIRST "SUMMARY MESSAGE"

PP SS

ALL

SUM

Comments:

GMT of capsule clock reading (hr., min., sec.)
Capsule elapsed time (hr., min., sec.)
Retrofire time setting (hr., min., sec.)

Cabin Pressure (PSI)
Suit Inlet Pressure (PSI)
Cabin Air Temp. (Degrees F)

Coolant Quantity (Percent)
Suit Inlet Temp. (Degrees F)
O_2 Partial Pressure (PSI)

Pitch Attitude (Degrees)
Yaw Attitude (Degrees)
Roll Attitude (Degrees)

DC Volts
DC Amps
AC Volts (Fans)

AC Volts (ASCS)
Fuel Quantity Auto Percent

C-Band Interrogator Deviation(MCS)
C-Band Transponder Deviation (MCS)
S-Band Interrogator Deviation (MCS)

S-Band Transponder Deviation (MCS)
UHF Deviation Main (KCS)
UHF Deviations Standby (KCS)

HF Deviation (KCS)
T/M High Freq. Center Freq. Shift (KCS)
T/M Low Freq. Center Freq. Shift (KCS)

Figure 4.1 An original Mercury Post-Pass TTY summary message. Photo courtesy of Arnie Aldrich.

Longer Mercury flights required some readjustments to flight operations, and especially recovery of the astronaut in case of an emergency. Gemini would fly much longer durations, sometimes spending long periods over vast expanses of ocean with little spacecraft contact. In some cases, in much higher orbits, it was possible to maintain contact with certain sites for extended times. Apollo would fly a quarter of a million miles to the Moon, and transmit almost too much data with continuous communication with the MCC all the way there and back. How would the flight control teams get the data they needed in order to monitor and control much more complicated missions flying even more complex spacecraft and launch vehicles? How would the new control center and tracking networks obtain and process the data for display? What would the flight controllers in the new control center need for computation and communications? What would the displays look like?

What would they need in order to display information from two astronauts, or even two spacecraft during a rendezvous? How were they to compute and display requisite maneuvers? What would they need in docking a Gemini spacecraft with its Agena target? A great many innovative operational concepts were needed for the new control center. How would they be finalized and influence the control center design and implementation?

4.1.1 The State of the Art Then

At any given point in time, a conceptual thinker considers what is available now. He knows the state-of-the-art and may have an idea of what is in the technology pipeline that he may or may not be able to use. When the new control center was conceived in 1962, the foresight of those thinkers was not all that far ahead. It is hard now, over half a century later, to appreciate how primitive such technology was at the time. Here are a few dates to place things into perspective for just one area of technology; communications.

Developments in Communications Technology

December 18, 1958	The top secret SCORE (Signal Communications by Orbital Relay Equipment) satellite was placed into orbit by an Atlas missile. It included the following broadcast: "This is the President (Eisenhower) of the United States speaking. Through the marvels of scientific advance, my voice is coming to you from a satellite circling in outer space. My message is a simple one: Through this unique means I convey to you and to all mankind, America's wish for peace on Earth and goodwill toward men everywhere."
August 12, 1960	The Echo 1 passive satellite was launched.
July 10, 1962	Telstar 1 was launched. It successfully relayed TV pictures, telephone calls, telegraph images, and the first "live" transatlantic TV link-up.
February 19, 1963	Syncom 1 (for "synchronous communication satellite") made by Hughes failed to reach geosynchronous orbit, but attained low Earth orbit.
July 26, 1963	Syncom 2 became the world's first geosynchronous (albeit not geostationary) communications satellite. President Kennedy telephoned the Nigerian Prime Minister aboard the USNS *Kingsport* docked in Lagos Harbor, in the first live two-way telephone call to be made between heads of government by satellite.
August 19, 1964	Syncom 3 was the world's first geostationary satellite and was able to broadcast the Olympics from Tokyo.

These examples show how apparent it was at the time, that flight operations would still have to rely on the Goddard Space Flight Center (GSFC) to improve the communications from the remote tracking stations to the new control center via terrestrial means. Soon after the move to Houston, engineers from Goddard met with their MSC counterparts to discuss the technical requirements that they would like to implement for the Gemini space tracking and data network.

These included:

- Unification of all command, telemetry, and radio signals from the Gemini spacecraft onto a single carrier frequency.
- Conversion from analog to the newer and much more bandwidth efficient Pulse Code Modulation (PCM) digital telemetry.

- Two acquisition aids at each tracking station (one for the Gemini and the other for the Agena target or other spacecraft) and the ability to slave the radar to either vehicle.
- Addition of computers at the network stations to facilitate the processing of both command uplink and telemetry downlink.

Some of these suggestions were thought to be too great a change from what was used during Mercury. Most of the Goddard proposals were first rejected. In fact, the discussion went on for another year! Some of the post-Mercury mission analysis found that certain operational concepts were not as necessary as initially presumed. For example, communications with the astronaut while over a remote site tracking station was better than first conceived. Astronaut performance was also better than first conceived, now that his performance in weightlessness was no longer a problem and the potential health issues that flight surgeons had been concerned about were not a problem. So perhaps not as many stations would be required, and in future NASA would not need to send as many flight controllers out to the remote sites as previously required for Mercury, especially those with high speed links to the MCC. However, modifications continued at many of the remote sites to support the new spacecraft. Wherever possible the point to point ground communications rates were increased, and the new digital systems were given software that allowed summary spacecraft data to be automatically sent to MCC at TTY speeds in the absence of high speed links.

4.1.2   New Computing Technologies and Concepts

The development of the computing architecture for Mercury started in an office building in Washington, DC. This effort was transferred to the Goddard Space Flight Center following its dedication on March 16, 1961. The architecture was primarily centered on flight dynamics operations, and the first two IBM 7090s were installed in Building 3 at Goddard. Three other IBM 709 (vacuum tube) computers rounded out the Mercury tracking network. One of these was at the Bermuda Control Center and tracking station, one at the Cape Canaveral Range Safety Officer's building, and one at the GE-Burroughs Atlas guidance building also at the Cape. While NASA had the hardware computing power, one NASA engineer, James Stokes, acknowledged, "We didn't know enough to specify the requirements for the software." As IBM did not know enough about spaceflight either, they went to Dr. Paul Herget, who in 1948 had written a book about orbit determination and was actively pursuing spacecraft tracking in the early 1960s. Over the next couple of years, NASA and IBM jointly created the software that was needed for Gemini using the new hardware technology that was becoming available during that period.

For Project Mercury, the operational concepts that drove the initial software design at Goddard had included interfacing the Mercury Space Flight Network (MSFN) with the Mercury Control Center at the Cape. This included processing radar data from the remote tracking sites and transmitting the low speed teletype messages back and forth. In addition to powered flight trajectory construction, it also required providing orbit determination and reentry trajectory computations. The outputs included altitude, velocity, flight path angle and abort impact points for launch and capsule attitude, retrofire times and splashdown points for reentry. In powered flight the main Goddard computer, the Cape Range Safety Officer's impact prediction computer, and the Bermuda Control Center computer would each give GO/NO-GO recommendations to the Mercury Control Center's Flight Dynamic's Officer. When an astronaut in orbit passed over a tracking station he would be provided updated retrofire times. This trajectory management concept would continue during Gemini, with Goddard backing up Houston. For Apollo, there would be no need for the Goddard computing complex, as all the required capabilities would be proven and centered in Houston.

For Gemini, the Mercury configuration at Goddard continued to receive the radar data from the remote sites and to compute flight dynamics parameters for transmission to the Cape MCC. To process the digital telemetry downlink from Gemini, Goddard placed two UNIVAC 1218 computers at the primary remote sites. By that time, Houston had been persuaded to accept Goddard's proposed changes – but with the important stipulation that the new computers at network remote sites be used only for telemetry processing and not for commanding. As the reliability of this configuration increased and flight operations management gained confidence, they finally allowed the computers to handle both telemetry and commanding. At first they were in the dual redundant mode and eventually they were configured one for telemetry and one for command.

Figure 4.2 A UNIVAC 1218 computer. Photo Courtesy of Sperry Rand.

The transmission rates were state-of-the-art, at 50,000 bits per second to the MCC from the remote sites that were linked by land lines. In fact, NASA had to install more submarine cables and land lines to accommodate the increased data flow. These improvements greatly increased the real time data available both at the Cape MCC and the new Houston MCC. By this time, the MCC at the Cape was now called the Mission Control Center and was being adapted for the early Gemini missions, while the new Mission Control Center in Houston (referred to as MCC-H) was being outfitted with new computers and control rooms.

Figure 4.3 The Cape MCC modified in March 1964 for Gemini 3. Photo courtesy of NASA.

Flight controllers had learned their responsibilities using the equipment that was available in the Cape MCC, in particular analog displays like meters, strip chart recorders, and X-Y plotters. They were reluctant to adapt to advances in technology even though Gemini, in both the spacecraft and on the ground, was going digital. Thus the new MCC-H started out with strip chart recorders, X-Y plotboards, and even computer constructed meter displays on its console digital display system. These end devices assisted the flight controllers to acclimate to the new control center, but as soon as the controllers got used to digital plotting and digital spacecraft systems displays driven by the MCC computers the older devices were abandoned.

On January 19, 1965, the new MCC-H passively monitored Gemini 2 which was an unmanned suborbital test. By March 23 the new MCC-H was designated as backup to the modified MCC at the Cape (by then called the Kennedy Space Center) for the first manned Gemini orbital mission with Gus Grissom and John Young. The Goddard computers were still prime for flight dynamics calculations and the Cape had been augmented with the UNIVAC 1218 digital computers to process telemetry and command functions such as "fire retro rockets," "turn on telemetry transmitters" and "ring astro alarm." The first stations to transmit real time telemetry back to MCC-H were (in addition to the Cape Canaveral Station, CNV), Bermuda (BDA) and the downrange DOD sites of Grand Bahama Island (GBI), Grand Turk Island (GTI) and Antigua (ANT). The new MCC-H proved it was ready to support Gemini with all the new processing and display technology now within Building 30.

Gemini 4 was launched on June 3, 1965 and this time Goddard was backup for trajectory computing, MCC-KSC was backup for the MCC-H, and the new Real Time Computer Complex (RTCC) on the first floor of Building 30 was prime, as was the third floor Mission Operations Control Room (now designated MOCR 2). After this, Goddard went ahead with their expansion of the Manned Space Flight Network for Apollo with additional capabilities such as the Deep Space Network, the addition of the Unified S-Band tracking, telemetry and command system, and the construction of space tracking and DOD stations, five tracking ships, and up to a dozen Apollo Range Instrumentation (ARIA) aircraft, as well as the use of two Intelsat relay satellites.

4.1.3   New Operational Concepts

Even as early as 1962 the planning for the Gemini missions included rendezvous, EVA, docking, and multiple spacecraft and launch vehicles. The data supplied by these vehicles would be processed and displayed to show more information than the small number of parameters that were available to Mercury flight controllers. The concept of operations required new positions for the MOCR. For Gemini that would include a Booster Monitor specifically for the launch phases of the Titan II and the Atlas Agena. The concept of having an onboard computer and performing maneuvers in space created the Guidance Officer position whose duties would be even more complex for Apollo.

Because the spacecraft and launch vehicles were new and quite different from Mercury, the operational procedures would have to change. The countdown for a Titan II would require different monitoring from the MOCR by the new Booster Monitor position. Likewise, the Gemini spacecraft had two astronauts and many times the number of systems telemetry parameters. The addition of vehicles such as the Agena (or on one occasion, a second Gemini spacecraft) for a rendezvous operation required more flight controllers and more mission rules to cover all the possible situations. Integrated countdowns were developed to enable the MOCR to track all the launch vehicle activities at the Cape and at the remote sites. (One of my responsibilities as an Operations and Procedures Officer was to create the countdown that integrated the countdown used by the Cape for the Titan II with the one used in the MOCR and the remote sites.)

While some thought was given to Apollo and the impact that this would have on the control center design, it was thought that the overall capability of the new MCC for Gemini could be modified to handle the Apollo missions. It was 1962, and Apollo was still years out, and there was an imperative to make progress in building and outfitting the MCC. The initial design concept for the MOCRs (not its construction) was completed on September 19, 1962, a month prior to Wally Schirra's MA-8 flight. The first unmanned launch of a Gemini-Titan II was still some 17 months away, and that would be monitored at the modified Cape MCC. The operational requirements Philco-Ford's Western Development Laboratories had for the MOCR contract in early 1964 was what they would build. However, as soon as the MCC-H requirements for the manned Gemini 3 mission had been specified, several engineers at NASA Mission Planning and Analysis (MPAD), supported by flight controllers, set out to develop the MCC-H requirements for Apollo. Especially important were the expanded lunar navigation and trajectory requirements and a new launch vehicle and two new spacecraft. It is a testimony to the MCC-H development contractors that the design of the flight controller consoles and the displays and controls were flexible enough to accommodate Apollo with mostly software modifications and several additional consoles.

4.2   A NEW ORGANIZATION

Only about 750 people from the Space Task Group (STG) at Langley, Virginia, relocated to Houston in 1962, but the new Manned Spacecraft Center was being organized as a new NASA Center for missions in the future. The old STG flight operations organization was still flying Mercury missions out of the Cape. It was clear that most of the Mercury people would be part of the new organization, but that they would have to regroup and train for Gemini and Apollo. Also the entire MSC was expanding greatly to accommodate all the new missions. Construction work started on the empty 1620 acres in the summer of 1962. By the end of that year the organization had reached 2,400 people, and others were being hired and moving in. By February 1964 a veritable campus had sprung up complete with trees, ponds and ducks. Spaceflight (not just flight operations) now required an organization much larger than had ever been envisaged during Project Mercury. While 8,000 NACA people joined NASA nationwide in 1958, the number had doubled by the end of 1960. The MSC would also grow further to accomplish its many missions and goals.

The MCC-H was just one of about 20 center-wide special facilities considered by MSC management, and therefore required a capable organization. In addition, management was aware of the need for program areas such as spacecraft project offices, spacecraft systems design and testing offices, other supporting roles such as mission design, contracting offices and public affairs. It was inevitable that the natural bureaucracy of government would begin to take hold. Long gone was the simplicity of the Project Mercury organization with its functional efficiency and the openness that enabled one to walk into someone's office to discuss an issue; even a senior manager. Soon there would be several layers of management from Washington on down. Even in "Operations" the organization was expanding. By this time it consisted of operational planning, crew selections and training, flight control, mission planning, experiments and simulation. A look at the September, 1962 MSC organization chart shows 52 "boxes" and only three have anything to do with flight operations: one for pre-flight operations at the Cape, one for control center operations, and one for the crew. Of course, another layer of management was above that in the form of what were called Assistant Directors.

4.3 A NEW APPROACH TO MISSION CONTROL

In the late 1950s, the approach to designing the Mercury Control Center was that a small group should get together to determine what "controlling" a capsule with a man in it was all about. The original concern was just to achieve a flight in one piece. Missile failures were quite common. Certainly things could happen too fast for the pilot (now called an astronaut) to escape the exploding missile. Back then, the concern was how the ground should help the astronaut once in orbit: What do the flight controllers know that the astronaut doesn't, and how should they assist him? Then the issue was how to ensure he was in the correct attitude for reentry? And what if the retrorockets don't fire? What did the flight control team need to monitor to help the astronaut to do what is necessary for the safety of the flight? And if the need arises, what should the ground itself control to ensure success of the mission? Support of flight experiments was always addressed last, because if the mission were lost, so too would be the results of the experiments.

Certainly, Gemini and Apollo would require a new and more comprehensive approach to mission control and flight operations, as well as a new approach to engineering of the control center. Outside of NASA flight operations there were companies possessing technical backgrounds that would be required. Some had technical and engineering backgrounds from the military, as well as those in the missile, aircraft, and computer and communications industries. Of course, there was the realization that the people who had manned the Mercury Control Center had unique insights into how the new mission control center should be designed.

While there were engineers building the NASA tracking network, and others building the military tracking stations, these were not mission control centers as envisioned for Gemini and Apollo spaceflights. Nevertheless, there were inputs to the new approach and physical similarities of apparatus and operations. So as Mercury ran through its final months, Chris Kraft and his flight controllers knew what they wanted for the new control center.

The fundamental approach included:

- It would be state-of-the-art.
- It would have a self-contained computer complex.
- It would use the Goddard complex for trajectory backup.
- It would use the NASA Manned Space Flight Network.
- It would have two identical control rooms to support multiple missions, training, and simulations.
- It would have staff support rooms for greater and faster depth of systems knowledge.
- It would have larger and more informative group displays.
- It would have computer generated digital console displays.

Kraft knew that when it came to procurement, not too many companies could support the planned missions, and he was satisfied with the support that IBM and Philco had given Project Mercury. A study contract went to Philco Corporations' Western Development Laboratories in April 1962 to study the flight information and control functions of the MCC. At that time, some flight control people were still flying Mercury missions out of the Cape and others were selling their homes in Virginia and moving to Houston. The STG move wasn't considered complete until July 1962.

Finding temporary office space to accommodate all the STG transferees and the new hires was a major challenge. The result was a hodge-podge of facilities; from offices to a bank to stores to apartment complexes, all located close to the Gulf Freeway around the Gulfgate Shopping Mall. They were a long way from the new site at MSC, where clearing and preparing the site for construction was underway. Most of the flight operations people were in the Houston Petroleum Center (HPC) and the adjacent Stahl Myers Building (a former sporting goods store). The human resources department for all the new employees and the STG transfers was in the East End State Bank building.

4.4   DESIGN

4.4.1   The Ground Based Computing System

The heart of the Mercury control network was an IBM 7090 mainframe computer located at Goddard. Its initial design was influenced by the Ballistic Missile Early Warning System (BMEWS). To provide the reliability needed for manned flights, the primary Mercury configuration included two 7090s operating in parallel, each receiving inputs, but with just one allowed to transmit output. These were named the Mission Operational Computer and the Dynamic Standby Computer, and the names stuck through the Apollo and Space Shuttle programs. This was NASA's "first" redundant ground based computer system. Switching over from the Prime to the Dynamic Standby was a human decision involving a single manual switch. The need for an active standby was proved during John Glenn's orbital mission, when the prime computer failed for 3 minutes. Three other computers completed the Mercury network. One was an IBM 709 dedicated to continuously predicting the impact points of missiles launched from Cape Canaveral. This provided data needed by the Range Safety Officer to decide whether to abort a mission during the powered flight phase and, if aborted, to supply information about the landing site for the recovery forces. Another 709 was at the Bermuda tracking station. It had the same responsibilities as the pair of 7090s at Goddard. It would become the prime mission computer in the event of either a communications failure or a double mainframe failure. The third of these other computers was the Burroughs-GE radio guidance computer which controlled the Atlas during its ascent to orbit. So Mercury had a total of five computers, all of them supporting flight dynamics and trajectory calculations. They all influenced the design of the computers and trajectory software for Gemini and Apollo. The fundamental concepts of backup and reliability remained constant throughout; only the implementation specifics changed from one program to the next. This design also impacted the operational decision process as failures of these components triggered Mission Rules and the GO/NO-GO decision process.

So we see that even before John Glenn flew his Mercury mission in February 1962, the Manned Spacecraft Center had been established and NASA engineers working on mission control were making inputs to the design of the new mission control center in Houston. Howard "Bill" Tindall worked on ground control for NASA from STG. He realized that basing the STG management at Langley, the computers and their programmers at Goddard, and the flight controllers at Cape Canaveral, would create serious communication and efficiency problems during Mercury. In January 1962, he began a memo campaign to consolidate all those functions and components at one site, obviously at the new Manned Spacecraft Center that had been officially designated just a few months earlier on October 24, 1961. On February 28, just 8 days after Glenn's mission, Tindall produced a detailed essay in which he pointed out that IBM was the only company with the capability to develop the necessary real time software. He also asserted that the Ground Systems Project Office, which had an oversight function for the MCC systems development, should permit representatives from the new MSC Flight Operations Division and Mission Planning and Analysis Division to assist with MCC design and software development. As the eventual users of the system, it made sense to include them.

Eighteen companies submitted bids for the RTCC, including RCA, Lockheed, North American Aviation, Computer Sciences Corporation, Hughes, TRW, and ITT. As MSC Chief Flight Director and head of the Flight Operations Division, Christopher C. Kraft, Jr. chaired the Source Evaluation Board (SEB) that studied the responses to the request for proposals. Tindall served too, with James Stroup, John P. Mayer, and Arthur Garrison – all of them at the new Manned Spacecraft Center.

On October 15, 1962, the SEB awarded the original computer contract for the "Ground-Based Computing System" to IBM for $36,200,018. The company had won the Gemini onboard computer contract for $26,600,000 back in April. This had significant software design and operations implications, since the computing system at MCC would routinely need to provide the onboard computer with new state vectors for navigational purposes and to process telemetry from the onboard computer.

The ground computer system contract was extended to December 1966. The total cost came to $46 million. With 6 weeks of preparation already behind them prior to the contract award, IBM's core team was in Houston by October 28 and ready to start work. So the design decision for the computing complex had been made. The facility would be called the Real Time Computing Complex (RTCC).

The design was based upon Goddard's Mercury experience but influenced by the rapidly moving state-of-the-art, in particular the move from analog to digital systems in the spacecraft and network. Mainframe computers of the era had just moved from the IBM 709 with vacuum tubes to the transistorized 7090s. It was Goddard's success with the IBM series of computers that led to the selection of the MSC operational computers.

4.4.2   The Mission Operations Control Rooms (MOCR)

The Flight Control Branch cooperated with other MSC organizations to include their inputs into the design process. Every flight controller had his own ideas on how the new control center should help to support his particular functions. This effort resulted in the production a document "The Flight Control Requirements for the Integrated Mission

Control Center." This was submitted to the winning study contractor. In April 1962 a Philco-Ford subsidiary, Western Development Laboratories (WDL) in Palo Alto, California began a study of the requirements for the new mission control center. One aspect of the study was to take numeric data and give it pictorial content, to make the jobs of the flight controllers more intuitive. Achieving this required much more sophisticated software and display equipment.

As Philco worked through the summer of 1962, NASA Administrator James Webb announced, on July 20, that there was to be an expanded replacement for Mercury Control. A request for proposals was prepared and sent to the industry. Seven corporations responded with proposals to build the second and third floor control rooms. Philco's design was broad in scope, covering physical facilities, information flow, consoles, displays, reliability studies, computers and software standards. It observed that modularity in program development would simplify maintenance and allow the use of one level of staff to code subroutines while a more experienced level developed the executive software. This organizational approach would become standard for large software program projects. Another requirement was that the probability of success in providing real time computer support for a 336-hour (14 day) mission must be 0.9995. Also, with rendezvous plans for Gemini and the dual-spacecraft Apollo lunar missions, the center must be capable of controlling two spacecraft simultaneously. To meet the reliability and processing goals, Philco investigated existing computer systems from IBM, UNIVAC, and Control Data Corporation, as well as its own Philco 211 and 212 computers to determine what type of computer and how many of them would be needed. The calculations resulted in three possible configurations: (1) five IBM 7094s, (2) nine UNIVAC 1107s, IBM 7090s, or Philco 211s; or (3) four Philco 212s or CDC 3600s. No matter which group was chosen, it was obvious that the complexity of the Gemini and Apollo computing facility would be much higher than its two-computer Mercury predecessor. To help keep the system as simple and inexpensive as possible, NASA specified to potential bidders that off-the-shelf hardware was essential.

4.5   BUILDING PHASES OF THE MANNED SPACECRAFT CENTER

One would think that the Mission Control Center was so important that it would be very high on the construction priority list, but it wasn't so. The Phase I work for the entire MSC started April 1962 and was completed on July 18, 1963. All the while, Mercury missions were being flown from the Cape. This construction task included preparing the site in terms of roads, utilities, and lighting.

In October 1962 the Phase II contract was awarded for the construction of the Data Processing Center (Building 12), the sewage plant, the central heating, the cooling plant, the fire station, the water treatment plant and associated buildings, all of which were finished by December 1963. Much of this work could be done in parallel with the Phase I work.

Phase III incorporated the largest grouping of buildings under one contract. On December 3, 1962, the contract was signed for eleven major facilities, including the project management building, the cafeteria, the (Building 30) flight operations and astronaut training facility, the crew systems laboratory, the technical services office and shop buildings, the systems evaluation laboratory, a spacecraft research lab and office building, and a data acquisition building and support buildings such as the shop building and warehouse.

In June 1963, Building 30 was finished sufficiently to start to outfit it with the systems it would hold. By April 1964 the RTCC was in place on the first floor. By January 1965 enough of Philco's third floor equipment was installed to allow the flight controllers to passively monitor Gemini 2. From that first use of the MCC-H it was only four months to Gemini 3, which was the first manned Gemini, even though the control center was just backup to the MCC at the Cape. Only a month after that, the MCC-H was prime for Gemini 4. Finally, the MCC-H was the fully operational and would remain so for almost three decades.

The Mission Operations Wing (MOW) of Building 30 is the equivalent of five stories high but has only three floors. Its first floor housed the Communications, Command and Telemetry System (CCATS) and its three UNIVAC 494 computers, the Voice Communications System, portions of the Display/Control System (which were also on the second and third floors) and the RTCC. The second and third floors contained the MOCRs, Staff Support Rooms (SSR), Simulation Control Room, Meteorological/Weather Room, a Recovery Control Room and other equipment rooms. It is the third floor MOCR 2, now renamed the Apollo Mission Control Center, that has been declared a National Historic Landmark and is the primary focus for this book.

The Operational Support Wing (OSW, later called the Administrative Wing) houses offices, a laboratory and technical support areas, a Mission Briefing and Observation Auditorium, an Auxiliary Computer Room, and the offices of the Flight Operations Directorate. The Lobby Wing interconnects the MOW and the OSW and contains several offices and (at the beginning of Gemini) a dormitory and cafeteria where certain critical flight controllers could be housed during the missions.

The new MCC-H, with all its support systems, computers, offices, equipment and unique consoles and displays, was ready to conduct spaceflight operations in a new age.

5

From Concepts to Reality

5.1 THE MISSION CONTROL CENTER-HOUSTON (MCC-H)

5.1.1 Background

The focus of this book is on the Apollo Mission Control Center. This is the term coined to identify the one control center that has been named a National Historic Landmark. But in the early days of the space program in Houston, the terms that we employed were the Integrated Mission Control Center (IMCC) and Mission Control Center-Houston (MCC-H), with the latter being standard once building work got started. The building housed two control centers and all the equipment to make them operational but it is only the third floor control center which is the National Historic Landmark (NHL). However, it was not the only national space facility to be considered for the honor.

The Man in Space National Historic Landmark Theme Study was prepared in 1984 by Dr. Harry A. Butowsky of the National Park Service. Although he listed a total of 24 facilities, he pointed out that these represented only a small fraction of the technological resources that were necessary to support the American space program. They were recommended for possible designation as National Historic Landmarks as the best and most important surviving examples of this technology. Due to the rapid change of the space program and evolving technologies, support facilities simply do not survive, or survive in a greatly altered state. The effort to land a man on the Moon, investigate the near Earth environment, and explore the solar system were supported from a technological base that reflected a depth and variety of support facilities that were unprecedented in American history. A great many of these resources have long since been destroyed, abandoned, or altered to satisfy the changing needs of the space program. The 24 facilities in Butowsky's study are only a fraction of this resource base. They are the best, most intact and most important resources that have survived. Fortunately, many of the facilities in the MCC-H have survived, with the most important being the Apollo Mission Control Center itself.

© Springer International Publishing AG, part of Springer Nature 2018   26
M. von Ehrenfried, *Apollo Mission Control*, Springer Praxis Books,
https://doi.org/10.1007/978-3-319-76684-3_5

5.1.2   The Original Layout

Between 1962 and 1963, Kaiser Engineers of Oakland, California, completed the design for the Mission Control Center (Building 30). As conceived, the MCC had an Administrative Wing (Wing A) and a Mission Operations Wing (Wing M) that was connected to it by a Lobby Wing (Wing L). While the Administrative Wing provided three floors of office space for the mission operations staff, as well as a "Mission Briefing and Observation Auditorium" on the first floor, the Missions Operations Wing included two Mission Operations Control Rooms (MOCR), a Recovery Control Room, a Simulation Control Room, many Staff Support Rooms (SSR) and various data and communications processing and equipment areas. In October 1962, IBM was selected to design the Ground Based Computing System, soon to be designated the Real Time Computer Complex (RTCC) for processing data from the spacecraft. In March 1963, the Philco-Ford Western Development Laboratories (WDL) in Palo Alto, California, was contracted to provide the other electronics, notably the communications center, the flight simulator facilities and the flight operations displays.

In late 1962, work on the MCC's foundations, structural steel frame, and roof frame was initiated by the joint venture of W.S. Bellows Construction and Peter Kiewit Sons, Company, both of Houston, Texas. Bidding for the other structural work was opened on March 15, 1963 and the contract was awarded to Ets-Hokin and Galvan, Inc., of Houston, Texas. The firm completed the construction of the facility between June 1963 and November 1964 under the direction of the U.S. Army Corps of Engineers (ACOE), for a cost of just over $8 million. In April 1964, IBM's RTCC was installed on the first floor of the Mission Operations Wing. On January 19, 1965 Philco had installed enough equipment in MOCR 2 for a team of controllers to passively monitor the Gemini 2 unmanned mission. The MCC served as a backup control center to the MCC at the Cape Canaveral Air Force Station, Florida, for the third Gemini mission launched on March 23, 1965. MCC-H officially took over all manned flight control operations with the launch of Gemini 4 on June 3, 1965.

Figure 5.1 The brand new MCC-H in early 1965. Note the new tree and the 1957 Ford on the far right. The Mission Operations Wing is on the left, the Lobby Wing in the center, and the Administrative Wing on the right. Photo courtesy of NASA.

After Gemini 4, the remaining eight Gemini flights were also controlled from the third floor MOCR 2. During this timeframe, the second floor MOCR 1 was configured to monitor the earliest test flights of the Apollo program, beginning with Apollo-Saturn 201 on February 22, 1966. After Gemini was wrapped up in November 1966, the third floor MOCR 2 received a year-long modification and reconfiguration in order to support the upcoming manned Apollo missions. The Apollo 1 test on January 27, 1967 was conducted from MOCR 1, being cut short by the capsule fire that killed the three astronauts. There was a long stand down before the first manned Apollo 7 flew from MOCR 1 in October 1968.

The first Apollo mission run from MOCR 2 was the unmanned Apollo 4, on November 9, 1967. Apollo 5, Apollo 7 and the Apollo-Soyuz Test Project were controlled from the second floor MOCR 1. The eight manned Apollo missions, Apollo 8 to Apollo 17 were all controlled from MOCR 2, so it was this control room that featured on televisions throughout that period, and this room that has become a National Historic Landmark.

5.1.3   Post-Apollo

After Apollo 11, the second floor MOCR 1 was deactivated and reconfigured for Skylab and Apollo-Soyuz Test Project missions. After the final lunar landing the third floor MOCR 2 was deactivated in May 1973 to enable it to be reconfigured for the Space Shuttle.

In June 1976, IBM was awarded a contract of roughly $24 million to provide the Space Shuttle Data Processing Complex (SSDPC) needed to handle all of the communications and telemetry data from the Shuttle vehicle. This included "the design, fabrication, delivery, installation and checkout of the computer complex and associated software." In effect, the SSDPC was to replace the RTCC which had served throughout the Gemini, Apollo and Skylab programs. The following month, Aeronutronic-Ford (previously Philco-Ford) received a contract "for the design, development, implementation, test, and maintenance and operation of the MCC-H covering the Shuttle Program, Design, Development, Test and Evaluation (DDT&E) period." This included the Orbital Flight Test Data Systems (OFTDS), the Shuttle Operation Data Systems (SODS), the Approach and Landing Test Data System (ALTDS), and the Shuttle Program Information Management Systems (SPIMS).

Installation of the ALTDS began in August 1976 and concluded in December 1976, with the Approach and Landing Tests scheduled to start in February 1977. The main function of the ALTDS was to enable flight control teams in the MCC to monitor the ALT missions. It was then integrated with the existing consoles in the third floor MOCR 2. In June and July 1977, the MCC successfully supported the ALT's three Captive-Active flights using the new software, and then the five Free-Flights from August through October. On November 9, 1977, the ALTDS was deactivated. By the end of June 1979 all of the old equipment was removed from the MOCR 2, to enable the room to be readied for the operational phase of the Shuttle.

5.1.4 The Space Shuttle Period

While flight controllers were monitoring the ALT missions, the first floor of the Operations Wing was reconfigured to support the orbital Shuttle. This process mainly involved the subdivision of larger rooms, or the enlargement of smaller rooms, into new spaces that were assigned to hold specific system components. Coinciding with this effort was the preparation of the second floor MOCR 1 to support the Orbital Flight Test (OFT) phase of the Shuttle program. It became a Flight Control Room (FCR); specifically FCR-1. If the reason for this change in terminology was to discard the "mouker" sound, then the "ficker" sound didn't accomplish much apart from drawing a distinction between Apollo and Shuttle! The logic was that they would be flying "flights" not "missions," therefore they would be called Flight Control Rooms or "fickers" for short.

The reconfiguration task included the removal of the existing consoles. Their frames were refurbished and fitted with new computers. The updated equipment, as well as the additional consoles for new flight control positions, were returned and arranged to support the Shuttle. Once this work was done, various interface tests between the SSDPC and the FCR equipment were conducted to ensure the system worked properly. In March 1979, FCR-1 successfully supported the first integrated simulation of the STS-1 mission. Afterwards, in-house and integrated simulations continued on a weekly basis, allowing the control teams to become better acquainted with their equipment and the systems of the new spacecraft. In November 1979, FCR-1 supported the first pre-launch testing at the launch pad. In January 1980, it participated in the first full simulation of a Shuttle mission in conjunction with the astronauts in their training simulators. This run lasted thirty hours. An even longer, fifty-four hour simulation was undertaken in April 1980.

Meanwhile, FCR-2 was officially reactivated. In addition to refurbishing the existing console frames, installing new computers, and arranging the facilities to meet the Shuttle program requirements, a secure operations system was installed because this FCR was to be the sole control room for all Department of Defense (DOD) missions. From April 12–14, 1981, FCR-1 successfully supported the first flight of the Space Transportation System (STS) program, STS-1. It also provided flight control for the remaining three OFT missions: STS-2 in November 1981, STS-3 in March 1982, and STS-4 in June and July 1982.

In September 1982, the first addition to the MCC was formally accepted. This was a visitor's lobby along the east side of the Mission Operations Wing. One of the security requirements for DOD missions was that there be an entirely separate entrance for JSC visitors. It had an elevator that led only to the second floor of the Mission Operations Wing, allowing tourists access solely to the Viewing Area of FCR-1. Two months later, FCR-2 was used to control STS-5, the first operational flight of the Shuttle program. From then through January 1986, FCR-1 controlled six flights and FCR-2 controlled fourteen.

After the loss of Challenger in January 1986, significant changes were made to the MCC, both in terms of construction and new equipment. In 1987 it received a second addition. This 3,472 square foot extension to the north end of the Mission Operations Wing was to provide additional mechanical rooms and storage spaces. The same year, the Mission Evaluation Room (MER) relocated from Building 45 to the third floor of the Lobby Wing in order to satisfy DOD requirements.

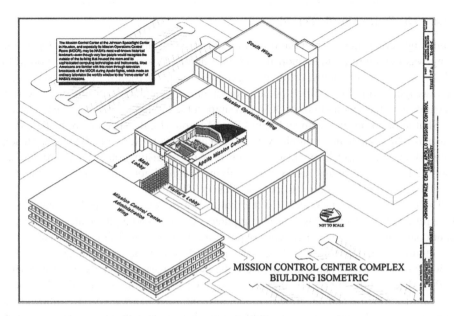

Figure 5.2   Additions to the Original MCC in 1982–1998. Photo courtesy of NASA.

In 1989, the largest addition to the facility, the Station Operations Wing, was built around the southwest corner of the Mission Operations Wing. This became operational in 1992. In 1998 a three-story mechanical room addition was located along the west elevation of the Mission Operations Wing. Over the years, several operational support rooms were created inside the existing wings but there was no additional major construction in the complex.

5.2   MISSION OPERATIONS CONTROL ROOMS (MOCR)

The Mission Operations Wing of Building 30 consists of three floors (although it stands five stories high) of a very large building some 168 by 224 feet. That is as wide as a football field and nearly as long, giving it a foot print of approximately 0.86 acres.

The first floor contained both the Real Time Computer Complex (RTCC) and the Communications, Command and Telemetry System (CCATS) that fed data to the RTCC to enable it to drive the displays of the flight controllers in the MOCRs. The first floor also contained voice, TTY, and TV communications equipment as well as maintenance areas for the enormous amount of equipment in the building.

This section focuses on the second and third floors that contain the MOCRs. The second floor contains MOCR 1 and the third floor contains MOCR 2, with the latter being the National Historic Landmark now called the Apollo Mission Control Center. The MOCRs aren't identical, and they changed roles as well as configurations to support Gemini, Apollo, Apollo-Soyuz, Skylab and the Space Shuttle.

The whole building complex changed again following the cancellation of the Shuttle, and the operations shifted to the International Space Station. This book will focus on the period which culminated in the Apollo lunar landing missions.

5.2.1 MOCR 1

The first thing you might notice is that the actual control room, like the tip of an iceberg, is only about 2,940 square feet; less than 8 per cent of the overall second floor area of 37,632 feet. The room #231 is only about 46 feet wide (not counting the hallway that goes by the MOCR) and is 64 feet from the center front screen to the glass wall behind the upper row of consoles. The ceiling stands 17 feet above the main floor. Surrounding the MOCR are rooms that support its operations, the display equipment installed behind the large wall screens, and spaces assigned to functional support areas such the Simulation Control Room, Staff Support Rooms and the Visitor Viewing Area behind the flight controllers that allows visitors an overview of the MOCR through a large glass wall.

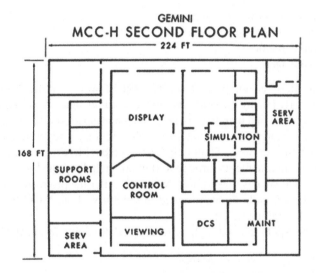

Figure 5.3 The second floor including MOCR 1. Drawing courtesy of NASA.

The flight controller consoles varied per mission. The Apollo 1 test that led to the fatal fire was monitored from MOCR 1. Only two manned Apollo missions were flown from MOCR 1: Apollo 7 and Apollo-Soyuz. For Apollo 7, the first manned mission, the positions (viewed from left to right as you look at the large screen displays) were as follows:

Front Row

• Booster Systems Engineer.
• Retrofire Officer.

- Flight Dynamics Officer.
- Guidance Officer.

Second Row

- Life Systems Officer (Flight Surgeon).
- CapCom (astronaut).
- CSM EECOM Engineer.
- CSM GNC Engineer.

Third Row

- Operations & Procedures Officer.
- Assistant Flight Director.
- Flight Director.
- Network Controller.

Fourth Row

- Public Affairs Officer/Commentator.
- Director of Flight Operations.
- Mission Director.
- DOD Manager.

Although this floor is not part of the restoration work for the National Historic Landmark, it supported many missions, including four unmanned Apollo flights launched on the Saturn IB and the first manned Apollo 7 and the manned Apollo-Soyuz Test Project, both of which were launched using Saturn IBs. In addition, it supported all four Skylab missions and the first four Space Shuttle flights before the MOCR was remodeled. It then supported six more Shuttle flights through to (but not including) the loss of Challenger on January 28, 1986, which was flown out of MOCR 2. After the protracted stand down following that accident, Shuttle flights resumed in 1988 from both MOCRs, with 41 being flown from MOCR 1 through 1996.

Specifically, the Apollo related flights (by launch date) flown from MOCR 1 were:

• AS-201	February 26, 1966.
• AS-203	July 5, 1966.
• AS-202	August 25, 1966.
• Apollo 5	January 22, 1968.
• Apollo 7	October 11, 1968.

Console positions were modified for Skylab:

• Skylab 1	May 4, 1973.
• Skylab 2	June 22, 1973.
• Skylab 3	July 28, 1973.
• Skylab 4	November 16, 1973.

The final Apollo mission was the joint project with the Soviet Union:

• Apollo-Soyuz	July 15, 1975.

Figure 5.4 Flight Director Glynn Lunney during Apollo 7, which was flown out of MOCR 1. Photo courtesy of NASA.

Figure 5.5 MOCR 1 during the Apollo-Soyuz mission in July 1975. Flight Director Neil Hutchinson is foreground-right. Alexey Leonov and Tom Stafford are on the video screen. Photo courtesy of NASA.

After Apollo, major changes were made to Building 30, including the RTCC as well as the MOCRs. Console positions were also modified to support the Space as follows:

•	STS-1	April 12, 1981.
•	STS-2	November 12, 1981.
•	STS-3	March 22, 1982.
•	STS-4	June 27, 1982.

Note that in 1984 the STS numbering system changed:

•	STS-41D	August 30, 1984.
•	STS-51A	November 8, 1984.
•	STS-51B	April 29, 1985.
•	STS-51F	July 29, 1985.
•	STS-61A	October 30, 1985.
•	STS-61C	January 12, 1986.
•	STS-26R	September 29/1988.

The flights that supported the International Space Station (ISS) after this date are beyond the scope of this book, which focuses primarily on those flights that were controlled from the Apollo Control Center MOCR 2; the last of which was STS-53 launched on December 2, 1992.

5.2.2   MOCR 2

The third floor MOCR 2, room #330, has the same dimensions as its counterpart on the floor below. It and four of the surrounding rooms are the only portions of Building 30 that have been designated the National Historic Landmark. The other areas included in the designation are the Display Projection Room, the Simulation Control Room, the Recovery Control Room, and the Visitor Viewing Area. These were included because they adjoin the MOCR, the iconic room at the heart of the nation's space program. These are now called the Apollo Mission Control Center and will be able to be viewed by visitors located in the Visitor Viewing Room. Actual access to the rooms will be limited and tightly controlled. In the case of the Recovery Control Room, the current plan is to possibly fix a large photograph to the glass window, because funds are not available to fully restore it. While the consoles in the Simulation Control Room are currently being restored, the room also awaits further funding before it can be completed.

Since the restoration focuses on the lunar landing missions, the following flight controller positions are those of Apollo 15 which, with few exceptions, represents all the Apollo lunar landing missions:

Front Row (Same as Gemini)

- Booster Systems Engineer (except for the Saturn).
- Retrofire Officer.
- Flight Dynamics Officer.
- Guidance Officer.

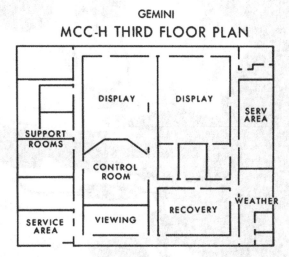

Figure 5.6 The third floor including MOCR 2. Photo courtesy of NASA.

Figure 5.7 One of the first Gemini flight control teams in the new MOCR 2 in 1965. Notice Neil Armstrong and Buzz Aldrin on the left. Gene Kranz is seated and yours truly is in the middle. Photo courtesy of NASA.

Figure 5.8 Flight Director Chris Kraft during Gemini 4 in June 1965. In the background are Bill Platt and Charles Harlan. Photo courtesy of NASA.

Figure 5.9 The new Gemini 4 MOCR 2 in June 1965. Gene Kranz, Chris Kraft, and John Hodge are at the Flight Director's console. This was probably during a shift handover. Photo courtesy of NASA.

Figure 5.10 MOCR 2 during Gemini 5. Front row, L-R: Tom Carter, Dave Massaro, Jerry Bostick, Charlie Parker, Will Fenner. Second row, L-R: John Aaron, Gerry Griffin, Arnie Aldrich. The man standing by the five Flight Dynamics Plotboards is not identified. Photo courtesy of NASA.

Figure 5.11 MOCR 2 during Gemini 6-A. Yours truly is on the Assistant Flight Director's Console. Photo courtesy of NASA.

Second Row (Addition of two Lunar Module positions)

- Life Systems Officer (Flight Surgeon).
- CapCom (astronaut).
- CSM EECOM Engineer.
- CSM GNC Engineer.
- LM EECOM Engineer.
- LM GNC Engineer.

Third Row (Addition of a Flight Activities & Experiments Officer)

- Operations & Procedures Officer.
- Assistant Flight Director.
- Flight Director.
- Flight Activities & Experiments Officer.
- Network Controller.

Fourth Row (No change from Gemini)

- Public Affairs Officer/Commentator.
- Director of Flight Operations.
- Mission Director.
- DOD Manager.

MOCR 2 supported the following missions between 1965 and 1982 (in terms of launch date):

•	Gemini 2	January 19, 1965 (Monitored for familiarization).
•	Gemini 3	May 23, 1965 (Backup to Cape MCC).
•	Gemini 4	June 3, 1965 (First Prime flight from MOCR 2).
•	Gemini 5	August 21, 1965.
•	Gemini 6	December 15, 1965.
•	Gemini 7	December 4, 1965 (Yes, ahead of Gemini 6).
•	Gemini 8	March 15, 1966.
•	Gemini 9	June 3, 1966.
•	Gemini 10	July 18, 1966.
•	Gemini 11	September 12, 1966.
•	Gemini 12	November 11, 1966.
•	Apollo 4	November 9, 1967.
•	Apollo 6	April 4, 1968.
•	Apollo 8	December 21, 1968.
•	Apollo 9	March 3, 1969.
•	Apollo 10	May 18, 1969.
•	Apollo 11	July 16, 1969.
•	Apollo 12	November 14, 1969.
•	Apollo 13	April 11, 1970.
•	Apollo 14	January 31, 1971.
•	Apollo 15	July 26, 1971.
•	Apollo 16	April 16, 1972.
•	Apollo 17	December 6, 1972.

After Apollo 17, the third floor was deactivated in May 1973 and reconfigured for the Shuttle. Flights continued from MOCR 1. The next flight out of MOCR 2 was not until STS-5. Twenty-two further Shuttle flights through STS-53 in 1992 were flown from MOCR 2, then it was deactivated for the last time. It has sat idle for over a quarter of a century, gradually deteriorating! It certainly deserves to be restored to its Apollo glory days for posterity.

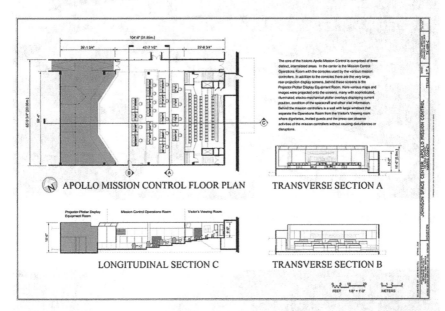

Figure 5.12 Floor plan of the Apollo Mission Control Center. The dark area is the Display Projection Room commonly called the "Bat Cave" because the walls, floors, and ceiling were all painted black. The text that is hard to read will be covered more thoroughly in the following sections. Photo courtesy of NASA.

5.3 STAFF SUPPORT ROOMS (SSR)

The need for additional support rooms became very evident during John Glenn's MA-6 flight in 1962. The "segment 51" issue left people in the Mercury Control Center scrambling for answers. There were two Mercury capsule manufacturer's representatives there to respond to questions; neither of them had good answers. Was the telemetry signal real? Did the landing bag deploy? Was the heat shield loose? Would Glenn burn up during reentry? The systems flight controllers had little telemetry data to tell them very much, other than to confirm that the signal was indeed present. Teletype messages were sent out to the remote sites for their input on what they saw from telemetry, and even what Glenn might have seen or heard. Ultimately the Flight Director Christopher Kraft and the capsule designer, Max Faget decided what to do.

In the future, the flight control management would make sure that they would have more depth of knowledge available. This led to the concept of having more people

Figure 5.13   MOCR 2 during the Apollo 11 mission. Photo courtesy of NASA.

Figure 5.14   MOCR 2 celebration after Apollo 11. L-R: Max Faget, George Trimble (partially hidden), Chris Kraft, George Low, Robert Gilruth, and Charles Mathews. Photo Courtesy of NASA.

Figure 5.15 MOCR 2 during an Apollo 12 EVA. Photo courtesy of NASA.

Figure 5.16 Three of the four Apollo 13 Flight Directors applaud the successful splashdown of the Command Module Odyssey while Dr. Robert R. Gilruth and Dr. Christopher C. Kraft Jr., light up cigars. The Flight Directors are L-R: Gerald D. Griffin, Eugene F. Kranz, and Glynn S. Lunney. Photo courtesy of NASA.

available to the flight controllers and the Flight Director, both in support areas at NASA and at the manufacturers. When the new MCC-H was designed, many rooms adjacent to the MOCR were designated for the different spacecraft disciplines. In fact there were many more people in these rooms than there were in the MOCR. And there were support rooms at the various contractors, such as North American Aviation and Grumman Aircraft, and support teams on hand at other NASA Centers. Depth of knowledge was now there "in spades."

Some of the rooms were called Staff Support Rooms (SSR), while others had their own names. Sometimes the rooms were called SSRs and other times called simply by their functions. During the lunar landing missions there was a need to support the Apollo Lunar Science Experiments Packages (ALSEP) as well as the experiments conducted on the CSM while in lunar orbit. For most of the Apollo lunar landing missions there were the following:

- *Flight Dynamics SSR*: Responsible to the Flight Dynamics Group located on the front row in the MOCR for detailed analysis of launch and reentry parameters, maneuver requirements, and orbital trajectories.
- *Vehicle Systems SSR*: Responsible to the Systems Operations Group in the MOCR for monitoring the detailed status and trends of the flight systems and detecting and isolating vehicle malfunctions. After deactivation of the S-IVB, the two booster consoles in the SSR were assigned to the Portable Life Support System (PLSS) engineer and the Experiments Officer.
- *Life Sciences SSR*: Responsible to the Flight Surgeon and accompanying aeromedical officers in the MOCR for providing detailed monitoring of the physiological and environmental data from the spacecraft concerning the flight crew and their environment.
- *Flight Directors' SSR*: Responsible to the Flight Director, Assistant Flight Director, Data Management Officer, and Flight Activities Officer. It was also responsible to the Apollo Communications Engineer for the detailed status of the communications system and two TV channels, as well as the Ground Timeline and Flight Plan.
- *Science/ALSEP SSR*: Responsible to the Experiments Officer, the Lunar Surface Program Office, and the Principal Investigators for undertaking detailed monitoring of the Apollo Lunar Science Package central station and experiments data. It was also responsible for all the scheduling of activities, commanding, and data distribution of the appropriate users.

5.3.1   Flight Dynamics SSR

The Flight Dynamics SSR (room #310) supported the Flight Dynamics Officer (FIDO), the Retrofire Officer (RETRO) and the Guidance Officer (GUIDANCE) in the first row of the MOCR, called the "Trench." Because of the complexity of lunar landing missions, this room was manned 24/7 and included a large number of people across all four operating shifts. Some people were needed for a specific phase (e.g. ascent) or maneuver (e.g. a midcourse burn) while others were needed for ongoing analysis.

In addition to NASA people, the following contractors were also represented in this room:

- MIT.
- IBM.
- TRW.
- North American Rockwell.
- Grumman Aerospace.
- McDonnell Douglas.
- Philco Ford.

The positions would typically include the following (with the floor layout as indicated):

- Maneuver Specialist (Position 1 Console 40).
- Trajectory Abort Chiefs (Position 2 & 3 Consoles 38 & 39).
- Apollo CSM Guidance Computer Support (Position 4 Console 37).
- Apollo LM Guidance Computer Support (Position 5 Console 36).
- FIDO, RETRO, GUIDANCE Support (Position 6 Console 35).
- Auxiliary Computer Room (ACR) Coordinator (With Consoles 2&3).

Most of the people there were representatives from the Mission Planning and Analysis Division, flight controllers and contractors. For a critical mission phase there could easily be twice as many people in this one SSR than there were in the MOCR.

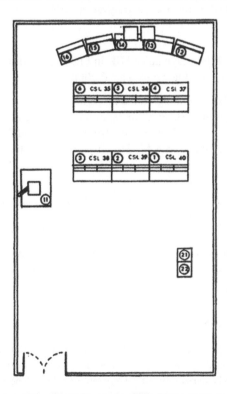

Figure 5.17 Floor layout of the Flight Dynamics SSR. Photo courtesy of Philco-Ford.

Figure 5.18 The Flight Dynamics SSR during Apollo. Not visible are the many X-Y plotboards in the front of the room. Photo courtesy of NASA.

The depiction of the Flight Dynamics SSR in Figure 5.18 shows some people who are now, half a century later, well known to spaceflight history:

- Howard W. Tindall (Standing on the far right).
- Lynwood C. Dunseith (Standing in the middle with the tan jacket).
- John R. "Jack" Garman (Seated in the front row; second from the left).
- Stephen G. Bales (Standing behind the front row with the dark jacket).
- John Jergeson.
- Bob Davis.
- Don Davis.
- Larry Davis.
- Ray Teague.

See Appendix 2 for the Mission Manning Lists of others who supported this SSR.

This room is not part of the National Historic Landmark and therefore will not be restored.

5.3.2   Vehicle Systems SSR

A spacecraft typically has several specialists trained in its systems, subsystems, and components. The MOCR flight controllers relied on their counterparts in the SSRs for expertise and in-depth analysis, especially in off-nominal or emergency situations.

For Apollo this SSR monitored and analyzed the Command & Service Module (CSM), the Lunar Module (LM) and the third stage of the Saturn launch vehicle (the S-IVB). The S-IVB engineers were called Booster System Engineers (BSE). Thus this SSR (room #311) would have both NASA engineers and those from the contractors who built those systems. They manned the room 24/7 with four shifts during long duration lunar flights.

The contractors included:

- North American Rockwell.
- Grumman Aircraft.
- IBM.
- McDonnell Douglas.
- Philco-Ford.

This SSR had the following positions:

- LM Guidance/Navigation/Control (GNC) Specialist (Console 29).
- LM Electrical/Communications Specialist (Console 30).
- CSM GNC Support Specialist (Console 31).
- CSM EECOM Support Specialist (Console 32).
- Booster Support Specialist (Console 33).
- Booster Support Specialist (Console 34).

Figure 5.19   The Vehicle Systems SSR. Photo courtesy of NASA.

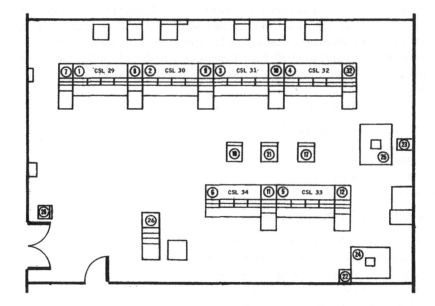

Figure 5.20 The floor layout of the Vehicle Systems SSR. Drawing courtesy Philco-Ford.

The other positions in the floor layout were locations for chart recorders, event recorders, and opaque televiewers. Counting all the personnel, this room would be occupied by as many as fifteen people on a shift. See Appendix 2 for the Mission Manning Lists of those who worked in this SSR.

This room is not part of the National Historic Landmark and therefore will not be restored.

5.3.3 Life Systems/Aeromed SSR

The Life Systems SSR (room #312), or the Aeromed SSR as it was more widely known, was manned by specialists concerned for the astronauts' health and well-being, particularly during high work-load activities such as EVA. As such it was staffed with people from the Life Sciences Division, the Crew Systems Division, and contractors who supplied the equipment that the astronauts were using. They were responsible to the Flight Surgeon in the MOCR and the other medical folks manning that console.

The SSR was staffed with the following positions:

- Surgeon (a medical doctor that managed the SSR activities).
- Data Engineer.
- Biomedical Engineer (and Aeromedical Representatives).
- Environment Control System engineer.
- EVA Support.

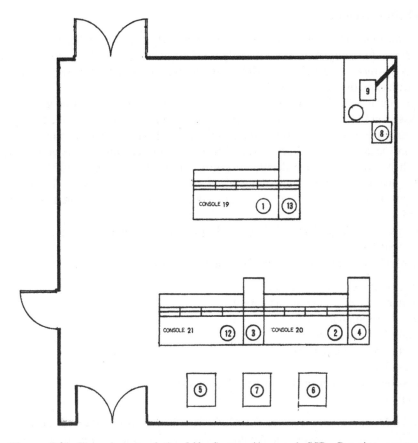

Figure 5.21 Floor layout of the Life Systems/Aeromed SSR. Drawing courtesy of Philco-Ford.

- Logistics Support.
- Clerical Support.

It was occupied by both NASA personnel and those of contractors responsible for the medical related equipment that the astronauts were wearing or using. This included the pressure suits and portable life support system equipment. This room would have up to a dozen people in it at any given time depending on the mission activity. Usually, they worked on a three shift basis, but often another shift was in for critical activities such as operating on the lunar surface, where the work-loads for the astronauts were high.

See Appendix 2 for the Mission Manning Lists of those who worked in this SSR.

This room is not part of the National Historic Landmark and therefore will not be restored.

5.3.4   Flight Directors' SSR

Support to the Flight Director evolved from Gemini to early Apollo, and then to the Apollo lunar landings. In reality there were literally hundreds of people in the SSRs and other support rooms ready to provide the MOCR flight controllers and ultimately the Flight Director with expertise at many levels.

This particular SSR initially provided support to the Operations & Procedures Officer and the Assistant Flight Director (and thereby to the Flight Director) for detailed information regarding the external support to the control center from the launch complex, from flight control teams at the remote sites, and from support personnel in the entire MCC building. It evolved to provide detailed expertise on all the mission requirements and objectives, operational procedures, flight plans, mission rules and many other support requirements. Eventually, the Flight Crew procedures team was incorporated, as well as CSM and LM procedures people.

What was originally an O&P SSR located in room #311 thereby became the Flight Directors' SSR for MOCR 2.

For Apollo 7, this SSR included the following positions:

- Mission Requirements Engineer.
- Requirements Configuration Control.
- Data Flow Support.
- Flight Crew Procedures.
- Flight Crew Experiments.
- Flight Plans.
- Mission Staff Engineer (Flight Test Objectives).
- Clerical Support.

I was the Mission Staff Engineer (MSE) for Apollo 7. I had worked for nearly two years on this mission following the Apollo 1 accident. This included working with the crew and the engineers to develop the flight test objectives that would be integrated into the Flight Plan and the Experiment Plan. I was also backup MSE to John Zarcaro for Apollo 8.

The lunar landing missions involved more experiments and longer EVAs, and the following positions were added to the SSR:

- Experiments Support.
- EVA Support.
- Mapping Sciences.
- CSM Procedures Support.
- LM Procedures Support.
- Photographic Tech Lab.
- Post-Evaluation Support.

These positions included both NASA and contractor personnel. There would be as many as 17 people in the room at one time, depending on mission activity. See Appendix 2 for the Mission Manning Lists.

This room is not part of the National Historic Landmark and therefore will not be restored.

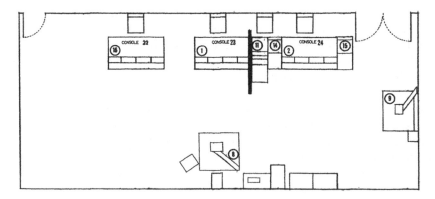

Figure 5.22 The Flight Directors' SSR. Drawing courtesy of NASA and Philco-Ford.

5.3.5 Science/ALSEP Support Rooms

In the years leading to the Apollo lunar landings, a consortium of distinguished scientists defined and prioritized a list of key scientific questions that were to be addressed by experiments conducted during space missions. Over several years, this work culminated in:

- A program that trained astronauts in the goals and methods of lunar field geology science.
- Developed the Apollo Lunar Surface Experiment Packages (ALSEP) that could be left on the Moon for at least one year to measure environmental conditions.

The experiment data would be radioed in real time back to the Manned Space Flight Network and then forwarded to the Mission Control Center. In effect, this hardware became lunar "weather stations" reporting to engineers, scientists, and technicians from NASA, other government agencies, universities, and a number of laboratories and support contractors.

Six packages were deployed on the lunar surface. They were completely self-contained science station that obtained scientific data and transmitted it to Earth, where the information was processed as part of the ALSEP support operations. Those operations were, for the most part, undertaken in the MCC in the ALSEP SSR, also referred to as the Satellite Control Room or the ALSEP Control Room. In addition to deploying the ALSEP, while on the lunar surface Apollo astronauts conducted field geology work that included collecting samples.

In lunar orbit, the Command & Service Module undertook a range of science experiments. For the final three missions, Apollo 15, 16 and 17, one bay on the Service Module included the Scientific Instrument Module (SIM) which housed cameras, spectrometers, and radiometers that recorded data in lunar orbit. EVA was required to retrieve film from the cameras during the voyage back to Earth. The Service Modules of Apollo 15 and 16 released a small package named the Particles and Fields Subsatellite (P&FS) into lunar orbit. Powered by six solar panels on its hexagonal sides and eleven silver-cadmium batteries, this satellite carried a magnetometer, an S-Band transponder, and charged particle detectors. Its task was to measure the strength and direction of the Earth's magnetic field in the vicinity of the Moon and the interplanetary magnetic field, to measure proton and electric flux from the solar wind, and to detect variations in the lunar gravity field. At such times, these rooms were shared with the scientists, engineers, and technicians who monitored the instruments of these satellites.

There are many aspects to monitoring and controlling science experiments in lunar orbit and on the lunar surface. Some experiments are conducted real time, but some of the ALSEP experiments ran on for years. These would initially be monitored 24/7 for months after being installed on the Moon, and then, after a certain time, only for a few hours a day. Some experiments required interaction with the MOCR flight controllers, others did not. There was a formal reporting and action item process. Activities that required a MOCR flight controller were managed by the SPAN. It was designated as the focal point to control and filter external inputs to the flight control team for specific support. In the MOCR the Flight Activities Officer was the primary interface to those in the SPAN who in turn handled the scientists' requests and/or the crews' requests for clarification about the deployment or operation of an experiment. And there was an Orbital Science Officer (OSO) who sat at the (vacant) Booster Systems console in the MOCR while science experiments were being undertaken. He was the interface between the flight control team, and the scientists and engineers in the Science Operations Room (sometimes called the Orbital Science Room) whose task was to monitor data from instruments orbiting the Moon.

The following rooms were involved in science monitoring and operations on Apollo flights. They were manned as necessary for the phase of the mission. The people and their positions are listed (consult the floor diagram with their relative positions):

- The Science Operations Room #314A included:

    1. Mission Scientist. Position 1 Console 88.
    2. Science & Applications Division representative. Position 2.
    3. Experiments Officer. Manager of the room. Position 3.
    4. Surface Operations Engineer. Position 4.
    5. Lunar Geology Experiment Principal Investigators. Position 5.
    6. Principal Investigators. Position 6.
    7. NASA Headquarters Monitor. Position 7.
    8. Science & Applications Division Data Manager. Position 8.
    9. Photography Coordinator. Position 9.
    10. Manned Spacecraft Center Chief Scientist. Position 10.

    This room was mainly responsible for the lunar geology experiments being conducted by the two astronauts on the surface. After the LM had left the Moon, additional scientists and engineers would move into this room to monitor the ALSEP.

- Satellite Control Room/ALSEP Room #314B (adjacent to #314A). This room included:

    1. ALSEP Senior Engineer. Manager of the room. Position 1.
    2. ALSEP Systems Engineer. Position 2.
    3. ALSEP Data Engineer. Position 3.
    4. ALSEP Principal Investigators. Position 4.

5. Plotters. Position 5.

This room supported ALSEP during the time that the six stations and their instruments were operating, running from July 1969 through to the termination of the program on September 30, 1977.

These adjacent rooms would be occupied by as many as 40 scientists, engineers, managers and technicians, depending on the level of activities. The scientists came from universities all over the country. The engineers were from the companies that made the experiments.

Both supporting teams were integrated into the overall MCC Mission Operations by including them in training and simulations, and further by coordinating their actions through an Experiments Officer in the MOCR using via voice loops and closed-circuit TV feeds.[1]

Over the course of the Apollo program the details of the integration of science into the MOCR discipline, which was strongly based on safety in flight, evolved as understanding and confidence between the two groups matured.

- MCC Science Data Room (#210A). This room on the second floor was a working area for the Lunar Geology Experiments Team during the lunar surface phase. These people coordinated with the Lunar Missions Office of the Science & Operations Division. There was also some coordination with the MER in Building 45 and with the Lunar Mapping Science Lab in Building 226.

These rooms are not part of the National Historic Landmark and therefore will not be restored.

The other support rooms, which weren't designated SSRs, are described in the next section.

5.4 OTHER SUPPORT ROOMS

Besides the SSRs (in the previous section) there were many other support areas that were conveniently located near the MOCR. They went by their functional names, as follows.

5.4.1 Recovery Operations Control Room

Because the recovery assets were almost all from the Department of Defense, in particular the Navy and the Air Force, they were managed by NASA by way of inter-agency agreements. The Recovery Operations Control Room (ROCR), in room #327, was managed by JSC employees but included representatives from the services. It was very large room with large screen displays run by projection equipment in an adjacent room to display the locations of all the recovery assets for planned and emergency situations. It was sited to the right of the entrance to the MOCR and behind a glass wall. It could not be seen by the

---

[1] Go to YouTube at https://www.youtube.com/watch?v=3Q8iP81GsuY for a three minute video of the team monitoring the Early Apollo Surface Experiment Package (EASEP) that was deployed by Apollo 11. Featured are Burt Sharpe, Franklin W. "Briz" Brizzolara, Donald J. McDonald (PHO), Dr. Gary V. Latham, Robert R. Miley and Rhea Q. Linney.

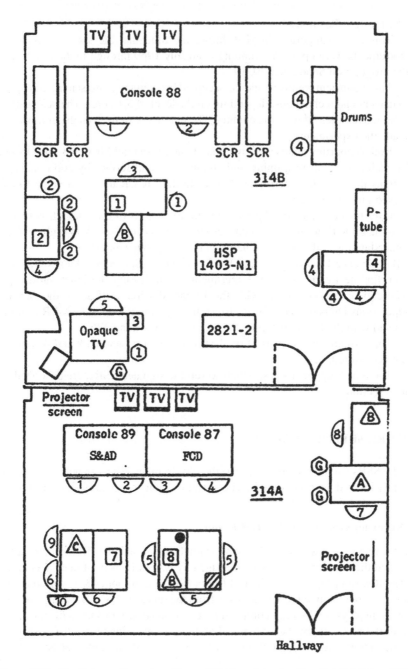

Figure 5.23 The Science Operations Room #314A and the Satellite Control Room/ALSEP Room #314B. The SCR annotation means Strip Chart Recorder, HSP means High Speed Printer, and Drums means Drum Data Recorders for long term experiment data for future analysis. A large glass window separated the rooms to ensure that the occupants of one room were aware of the level of activity in the other room. Drawing courtesy of NASA and Philco.

Visitor Viewing Room but could be seen from the DOD Manager position in the MOCR. He was on the back row, on the end facing the ROCR.

It was the responsibility of the Landing and Recovery Division (LRD), which was headed by Jerry Hammack. For the Apollo lunar missions, with only one or two exceptions, the ROCR NASA personnel came from this Division. There were five Branches in the LRD as follows:

- *Recovery Operations Branch Chief*: Dr. Donald E. Stullken.
- *Planning and Control Branch Chief*: Harold Granger.
- *Recovery Electronics Branch Chief*: William R. Chase.
- *Operations Test Branch Chief*: Weldon B. "Gus" McCown.
- *Recovery Systems Branch Chief*: John C. Stonesifer.

Harold Granger's Branch had primary responsibility for ROCR manning and operations. Section Head of the Control Center, Edward Bullock, was assigned the responsibility for operating the ROCR, with Donald Stullken's Branch (and others) supplementing the manning.

The Department of Defense Manned Spaceflight Support Office (DDMS) at Patrick AFB in Florida was responsible for providing military personnel to man the ROCR, as well as providing other support to Recovery Operations globally. Their MCC support was coordinated through the NASA Recovery Officer who had final authority for activities and decisions in the ROCR.

Recovery Operations involved the support of the DOD all around the world. This needed extensive ship and aircraft support, both in and around the Pacific and Atlantic. This further led to the requirement for a large number of NASA personnel being deployed as "advisers" to ships, aircraft, and Recover Control Centers. Consequently, the two Operations Branches could not support all the required positions. Other Systems Branches included the personnel who had designed, built, and tested much of the equipment that was used for recovery operations. These people were also utilized for deployments, in particular on aircraft and secondary ships where their equipment was employed. There was cross training on positions to enable people to be deployed wherever the need arose.

The ROCR was similar in layout to the MOCR. It was equipped with smaller forms of the large screen group displays. On the left was a Projection Plotboard to display the world map and ground tracks. In the center were two screens that were supported by vu-graph machines on earlier flights, then projectors on later flights. The projectors were used in conjunction with the overhead cameras (these can be seen in Positions 12 and 30). The central screens displayed the status of recovery forces and mission events. On the right was an Eidophor display that was used to present console systems displays or video feeds, such as real time KSC or MCC live links or downlinks from the crew. Above these screens were time displays to show GMT, Mission Elapsed Time, Flight Events, countdown clocks, etc.

Along the left wall was the P-Tube station. From the right of that was a long plot table with map files beneath it. The table extended for the length of the wall and was used for plotting and selecting target points.

In the back of the room was a smaller room (two rooms on the later missions) with the DOD Assistant for Communications, encryption equipment, and a TTY machine.

The ROCR had the following console positions:

- NASA Recovery Officer (representing the Flight Director in the MOCR) occupied the right side of Console 54 Position 1 in the layout depicted. He had final authority in the ROCR for all NASA recovery issues, including:

  1. Assuring the selection of the most desirable target points for any situation and determining the plan to best support medical, PAO, and experiment recovery requirements.
  2. Informing the DOD Operations Officer and NASA personnel of mission status and developments or plans which might affect the recovery forces.
  3. Informing the Flight Director of the status of the recovery force and weather conditions as they were likely to affect the recovery operations.
  4. Maintaining the NASA ROCR log.
  5. Preparing the Post-Mission Report.

- NASA Assistant Recovery Officer occupied the left side of Console 54 Position 1 during some mission phases, particularly launch and recovery. This person was a qualified Recovery Officer. The Chief of the Landing and Recovery Division would often sit here as well. At such times, the Assistant would move over to the left end of Console 55. The following responsibilities were involved:

  1. Taking the lead for interfacing with other ROCR personnel.
  2. Assuming the Recovery Officer's duties during his absence and assisting as necessary.
  3. Maintaining the Teletype board and console log.

- NASA Evaluator/Display Controller had Console 55 Position 3 with the following responsibilities:

  1. Assimilating and evaluating all data necessary to select the most desirable target points for any situation and recommending them to the Recovery Officer.
  2. Interfacing with the Flight Dynamics Officer and Weather Office about the target points.
  3. Determining the actual splash point.
  4. Assuming the duties of the Assistant Recovery Officer when this was necessary.
  5. Performing the duties of the Status Monitor when this position was not manned.

- Recovery Logistics Representative occupied Console 57H Position 4 with the following responsibilities:

  1. Working directly with the DOD Logistics representative and other NASA logistics personnel to arrange for the acquisition and movement of recovery-related equipment to designated locations.
  2. Integrating changes to the CM deactivation procedures (following recovery) and also procedures for the removal of equipment and experiments from the CM.

    3. Advising personnel external to the MCC of the crew, experiment, and CM logistics return schedules and transportation status.

    4. Scheduling equipment and support as necessary to receive crew samples upon arrival at Ellington AFB, near Houston.

- NASA Recovery Status Monitor occupied Console 57G Position 5. For earlier flights, he supported the entire mission, but subsequently he only supported the launch and reentry. He had the following responsibilities:

    1. Assembling and displaying on ROCR group displays, information on recovery force positions and status, pertinent recovery weather data and significant mission events.

    2. Preparing and presenting short mission status briefings for ROCR and deployed recovery personnel.

- DOD Operations Officer (representing the DOD Manager in the MOCR) occupied Console 56 Position 2. He was the DDMS representative to the DOD Manager for interfacing with NASA and for directing the activities of DOD personnel and forces assigned to the mission.
- DOD Primary Operations Officer relative to the Primary Landing Site (PLS) occupied Console 57D Position 9. This DDMS representative was responsible for assisting the Operations Officer.
- DOD Secondary Operations Officer relative to Secondary Landing Sites (SLS) occupied Console 57A Position 11. This DOD representative was responsible for assisting the DOD Operations Officer.
- DOD Coordinator occupied Console 57B Position 10. This was a DDMS representative to the DOD Operations Officer for communications with DOD personnel at the Recovery Control Centers (RCC).
- DOD Recorder occupied Console 57D Position 8. This was the DDMS representative responsible to the Operations Officer for logging all DOD activities.
- DOD Logistics Officer. The DDMS representative for directing recovery support logistics.
- DOD ARIA Coordinator. This DDMS representative reported directly to the DOD Manager and kept him advised of ARIA status; also kept other ARIA personnel advised of mission activities.
- DOD Public Information Officer occupied Console 57E Position 7. This DDMS representative was responsible to the DOD Manager for directing the DOD public releases.
- DOD Communications Coordinators occupied Console 57F Position 6 in a separate office at the back of the room. These DDMS representatives were responsible to the DOD Manager for assuring continuous support from the Recovery Communications Network.

Although there were eleven consoles in the room, they were often manned by additional personnel and secretarial support.

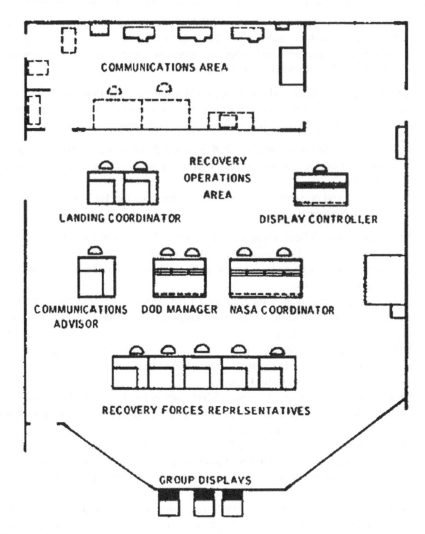

Figure 5.24 Floor layout of the Recovery Control Room. Drawing courtesy of NASA and Philco-Ford.

Sadly, due to insufficient funds this room is not part of the National Historic Landmark and therefore will not be restored. To provide an indication of what it contained, a large photograph will be attached to the glass window that faces the MOCR.[2]

---

[2] Go to YouTube and type in "Recovery Control Room at the Mission Control Center during Apollo 11" for a seven minute video.

Figure 5.25 The Recovery Operations Control Room during the Apollo 15 mission. The console in the foreground is the NASA Evaluator Console. On the left is Dick Snyder (also a Recovery Officer). To his right is John Hamlin, the Status Monitor. In the next row of consoles is the NASA Assistant Recovery Officer (not identified) and to his right is the NASA Recovery Officer, Chuck Filley. On the screen is a playback of the launch. Photo courtesy of NASA and Richard Snyder.

5.4.2 Spacecraft Analysis (SPAN) Room

The SPAN was the official interface for the Manager of the Apollo Spacecraft Program Office (ASPO) with the Flight Operations Directorate (FOD) and the Science & Applications Directorate (S&AD).

The ASPO had the capability to:

- Answer questions asked by the FOD prior to and during real time flight operations.
- Provide ASPO inputs, as required, on experiment hardware, spacecraft operations and mission requirements.
- Provide in depth, real time system performance analysis.
- Provide a means of drawing upon expert knowledge and the assistance of specialists (e.g. those in the Mission Evaluation Room).

The SPAN is located on the third floor in room #312A, adjacent to the MOCR. It is a mission support room (although sometimes called a SSR) connected to the MOCR through consoles and displays similar to the SSRs. The SPAN Operations Manager could quickly and accurately respond to questions posed by the MOCR flight controllers about the design

and operation of the spacecraft. It also enabled the ASPO staff to receive recommendations from the Mission Evaluation Room (located in Building 45 during Apollo but moved
to Building 30 for the Shuttle) and relay, if required, major decisions to the MOCR flight
controller who issued the request for support.

The SPAN was staffed by the following ASPO personnel:

- SPAN Operations Manager (the senior ASPO representative).
- Mission Planning and Analysis Division (MPAD) Senior Representative.
- Marshall Space Flight Center (MSFC) Lunar Roving Vehicle (LRV) Senior
  Representative.
- Mission Staff Engineer for Detailed Flight Objectives.
- Contractor Representative (NR/GAC/MIT).
- Crew Systems Division (CSD) Senior Representative.
- Log Manager for action items.
- Messenger to hand carry actions and responses to other rooms.
- Operations Manager.
- Secretary.

The SPAN was also staffed by the following FOD personnel:

- FOD Senior Representative.
- Flight Control Division (FCD) Senior CSM Representative.
- FCD Senior LM Representative.
- FCD Senior Flight Dynamics Representative.

It was also staffed by the S&AD Chief, Science Mission Support Division, to coordi-
nate with personnel, in two other rooms, of science teams for experiments such as the
Scientific Instrumentation Module (SIM) of the CSM and the Apollo Lunar Science
Equipment Package (ALSEP).

There was a formal process for requesting an evaluation and then responding to a
request. A request was logged in and assigned to the appropriate area by the Operations
Manager. Sometimes this action would involve people from both the NASA engineers and
the relevant contractor. The answer would be signed off by those responsible and the
answer documented using a SPAN Mission Evaluation Action Request (SMEAR). The
request would be formally logged out when the action was completed.

See Appendix 2 for the Mission Manning Lists of those who worked in the SPAN during Apollo.

This room is not part of the National Historic Landmark and therefore will not be
restored.

5.4.3   Mission Evaluation Room (MER)

The concept of a team of specialists who would help solve mission problems in "semi-real
time" was prompted by the troubles of John Glenn's mission. It was better defined while
supporting the Gemini program. A mission evaluation plan published in August 1968 formally established the mission evaluation team that provided engineering and technical
support to the MCC. The initial idea was to provide technical support by means of verbal

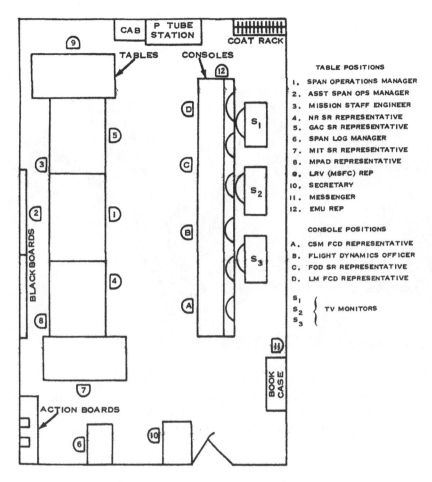

Figure 5.26 Floor layout of the SPAN room. Photo courtesy of NASA.

inputs to MCC personnel. Indeed, during the Apollo unmanned flights the engineering support sat alongside their flight control counterpart at a console in the Vehicle Systems SSR. But Flight Operations realized the SSR was getting too crowded, and would require even more people to support the later manned missions. After the unmanned Apollo test flights, the concept was revised to require written requests to the evaluation team and responses to the ASPO Manager's representative, whose place was in the Spacecraft Analysis (SPAN) room.

More detailed plans and assignments were documented for the Apollo 4 to 7 missions. Starting with Apollo 7, the first manned mission, the engineering and program management support, both NASA and contractor, was installed on the third floor of Building 45 and called the Mission Evaluation Room.

The original rationale for having a mission evaluation plan, was to provide the assignments and to define the responsibilities for personnel participating in the in-flight and

Figure 5.27 The SPAN room during Apollo 11. Seated at the table in the foreground are George Merrick from Rockwell, Dale Myers, Ron Kubicki, and Tom Kelly from Grumman. Standing against the wall are Aaron Cohen (middle) and ASPO Manager Owen Maynard (to Cohen's left). The man to Cohen's right is unidentified. Standing in front of the book case is Jim Hannigan. Those against the wall to the right on the consoles are in the next figure. Photo Courtesy of NASA.

post-flight mission evaluation efforts, but it was soon realized that other mission related information such as post-mission reports, procedures and schedule requirements should be added to the effort. As the Apollo missions became more complex during pre-launch preparations and in flight, the team had to provide real time support to both the pre-launch and experiment checkout activities at KSC in addition to the flight controllers in the MOCR.

A group of specialists was assigned to each engineering discipline necessary for a mission evaluation. Working in the Mission Evaluation Room (MER), the evaluation team provided support 24/7 during the mission, using three shifts of personnel organized under individual shift team leaders. All teams reported to a NASA shift manager – also known as the team leader – who was responsible to the evaluation team manager. They were both members of ASPO. The specific disciplines represented for each Command & Service Module (CSM) and Lunar Module (LM) spacecraft were its telecommunications, crew systems, electronic systems, propulsion and power, guidance and control, structures and mechanics, and thermal control.

In addition, there were specialists for the Apollo Lunar Surface Experiments Package (ALSEP), and safety, reliability, and quality assurance and flight crew training. Each shift

Figure 5.28 SPAN personnel on the consoles during Apollo 11. Seated on the right is Bill Blair from Rockwell, Arnie Aldrich, Joe Roach, and others not identified. Not in the photo are SPAN Manager Scott H. Simpkinson and Team Manager Donald D. Arabian. Photo courtesy of NASA.

included NASA personnel and contractors, and worked as a unit under a NASA team leader who directed overall efforts, resolved problems, scheduled evaluation tasks to meet time constraints, coordinated with other team leaders to ensure that the resolutions or recommended actions did not jeopardize other activities, approved the systems evaluations, and wrote the daily summary reports.

Corresponding teams of specialists were based in a mission support room at each of the two spacecraft contractor facilities: North American Rockwell and Grumman Aircraft. The work of each support team was coordinated through a contractor senior engineering manager assigned to the mission evaluation team who worked directly with the shift manager and the evaluation team manager.

Apollo 13 resulted in the Lunar Module (LM) being operated far outside its operational design limits. This meant that the MER NASA and contractor LM engineers, and managers and personnel at the manufacturer Grumman were all deeply involved in supplying information to the flight control team. The NASA environmental control team developed the procedures for the crew to create an adapter to allow the Command Module lithium hydroxide cartridges to operate with the LM environmental control system in order to reduce the high levels of carbon dioxide. And MER electrical engineers developed

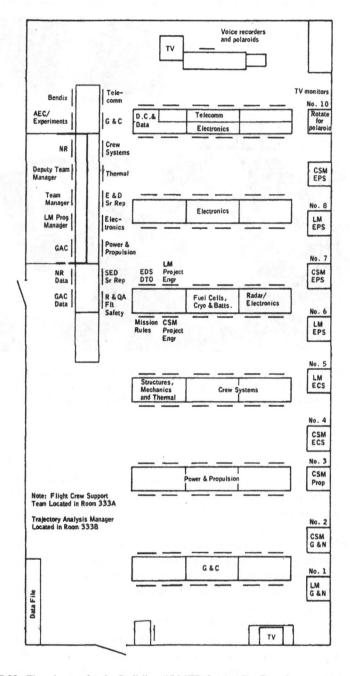

Figure 5.29 Floor layout for the Building 45 MER for Apollo. Drawing courtesy of NASA.

Figure 5.30 The Building 45 MER during Apollo 11. Photo courtesy of NASA and Gary Johnson.

a procedure that was supplied to the flight control team to allow a power transfer from the LM to the CM to prepare that vehicle for atmospheric entry.

In the wake of the aborted Apollo 13 mission, major changes to the mission evaluation team were instituted. An investigation team recommended that MSC subsystem personnel help resolve launch-site vehicle checkout problems, so the scope of the team was broadened to include pre-launch surveillance. Additional requirements and disciplines for pre-launch checkout were added to the mission evaluation plan. Consequently, the period of responsibility included continuous coverage from the beginning of the integrated systems tests at the launch pad to mission termination. Starting with Apollo 15, the team was enlarged to include specialists for the Scientific Instrument Module (SIM) bay (generally known as the "Simbay") of the CSM and the Lunar Roving Vehicle (LRV) that was to be delivered to the lunar surface by the LM. The mission evaluation team, initially located in Building 45, interfaced with the MCC through the ASPO Manager or his designated representative in the spacecraft analysis (SPAN) room.

Also in the SPAN were Flight Operations Division (FOD) representatives for each of the two spacecraft: CSM and LM. For example, the CSM representative had a console position in addition to a senior FOD representative. All of the FOD vehicle positions went to senior FOD management personnel (i.e., Branch Chiefs) and were badged for the MOCR to permit face-to-face interface with the MOCR flight controllers if required. These vehicle representatives provided the interface between the MOCR flight controllers and the ASPO and MER team. The SPAN developed the requirements for many of the mission evaluation team taskings.

The SPAN/Mission Evaluation Action Request (SMEAR) process (informally referred to as a "CHIT") was very formal. It could be initiated by a MOCR flight controller, the SPAN FOD representative, or the ASPO Manager. Or there could be requests going the other way. The MER often submitted formal requests to the MOCR. They were all handled the same way. There was a formal documentation procedure to permit tracking and approval.

Requests produced a disciplined effort, and facilitated an ASPO management review prior to sending the response. When responses had to be expedited, these were given verbally to the ASPO Manager or his representative and then logged in. If necessary, an action request form was prepared after the fact. In addition to this process, the team provided periodic systems reports and a daily report. Both reports were circulated at the appropriate levels of Center management. And the team manager briefed the ASPO Program Manager (George Low) on significant areas of concern before major mission milestones. The mission evaluation team responded to approximately one request for each hour of elapsed mission time. Some requests could take a considerable amount of time to develop a solution or answer, and therefore were not usually completed in real time or even near-real time. Some answers could take hours to days, or in some cases not until a post-flight analysis was conducted.

The following examples are representative of the many unexpected problems that occurred during Apollo missions, and how they were resolved:

- When the cryogenic oxygen supply was lost as Apollo 13 was heading to the Moon, the mission evaluation team, via the SPAN room, became the focal point for providing procedures for using the LM as a lifeboat. As a result of its experience and training in the evaluation of novel problems, the team played a major role in the successful return of the astronauts to Earth, as, indeed, did the entire flight control team and all the SSRs and other organizations at MSC.
- After six unsuccessful attempts to achieve capture in order to extract the LM from the S-IVB, the crew of Apollo 14 was finally able to achieve a capture. This problem required resolution before the spacecraft could be committed to a lunar landing. During the transit phase to the Moon, the team developed special troubleshooting procedures for the astronauts to perform. The MER team supplied alternate methods of undocking before descent and developed a procedure to retract the probe. This allowed the crew to achieve docking by contacting and tripping the structural latches in case capture was unable to occur again to complete docking following lunar rendezvous. A complete briefing of these alternatives was presented to the MOCR Mission Director and Flight Director prior to committing to a lunar landing. Because the docking system later operated satisfactorily, the procedures developed to circumvent such docking problems were not required. To facilitate return of the docking probe for analysis, the team provided a procedure and established a location for stowing the probe in the CSM for entry and landing.
- The Apollo 15 crew reported the Service Propulsion System "Thrust On" light was illuminated during transposition and initial docking with their LM, although it clearly was not true. This was a NO-GO for Lunar Orbit Insertion (LOI). The MER CSM electrical engineer realized that a short to ground in the "Delta V Thrust"

panel switch would illuminate the "Thrust On" light without issuing a fire command to the SPS. The MER Manager gained approval to send the engineer into the MOCR to explain how this could have caused the problem. The engineer went over the analysis with the MOCR Mission Director, who stated that some form of proof would be required. As a result, the team was asked to appraise the situation and determine a safe way to undertake the LOI maneuver. A troubleshooting procedure was devised and then executed by the flight controllers in the MOCR working with the crew. The fault was identified as a short circuit on the downstream side of a switch. An alternate procedure was specified for operating the engine, thereby enabling the mission to be completed as planned.

The MER procedures and the support facilities used during Apollo provided a basis for supporting Skylab, ASTP, and the Space Shuttle. On October 12, 2004, the new Space Shuttle Program Mission Evaluation Room (MER) was officially unveiled. It was relocated from Building 45 to the first floor of Building 30, and supported both the KSC launch control team and the MCC flight control team.

Figure 5.31 The last MER team for the Apollo-Soyuz mission prior to the MER's relocation to Building 30. Photo courtesy of NASA and Gary Johnson.

The Building 30 Space Shuttle Program MER seats approximately 162 people supporting 39 different positions. Each console has chairs to support two people, but there are 60 folding chairs to permit a third or fourth person per console on a short term basis. Nowadays, a similar MER provides support to the International Space Station.

Figure 5.32 The new Building 30 MER. Photo courtesy of NASA.

5.4.4   Spaceflight/Meteorological Room (aka the Weather Room)

This room was responsible to the MOCR and other SSRs and support rooms for meteorological and space radiation information. It provided weather analysis to the Recovery Support Room for the launch and recovery phases. It also supplied information on the deep space radiation environment which had potential effects on the crew and so was of great interest to the MOCR flight surgeon and the Life Sciences SSR.

It was generally a 24/7 operation with three shifts and the following console positions:

- Meteorologist (NASA, Weather Bureau and ESSA representative).
- DoD Defense Meteorological Service Officer.
- Space Environment instrument specialist.
- Space Environment radiation dose specialist.
- Space Environment solar physics specialist.
- Space Environment Team Leader.
- Space Environment Aeromed.
- Space Environment Officer.
- Goddard Communications Representative.

Depending on the mission phase, it would be staffed with up to 15 people. See Appendix 2 for the Mission Manning Lists of those who served in this support room.

This room is not part of the National Historic Landmark and therefore will not be restored.

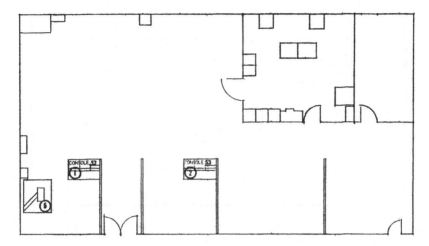

Figure 5.33 Floor layout of the Meteorological/Weather Room. Drawing courtesy of Philco-Ford.

5.4.5 Simulation Control Room

The flight controllers and all of the support personnel in the entire building and at all the locations around the world, spent many more hours simulating a flight than it took in real time. This type of training has paid off more than once, as shown by the safe return of the Apollo 13 astronauts.

More goes into simulations than meets the eye of one observing a flight from the Visitor Viewing Area or by watching it on TV. For the Apollo lunar missions this type of training was undertaken in room #328. There were console positions for the Command & Service Module (CSM), the Lunar Module (LM), the Saturn Launch Vehicle (SLV) and its S-IVB upper stage, as well as the communications and tracking network. These operators could simulate various types of failures to assess the reactions of the flight control teams. This needed computer simulation software that would drive the flight controllers' displays. Often, new operational procedures and mission rules were written in response to such simulations. Even the astronauts' checklist and procedures could be rewritten.

The Simulation Team faced the MOCR through a glass wall so they could see the reactions of the flight controllers. The second floor Simulation Control Room included more consoles for training and the equipment to drive the simulations.

Sadly, the third floor Simulation Room is currently a storage room and holds equipment other than Apollo. Because it is visible from the MOCR and partially from the Visitor Viewing Area, it will be partially restored to reflect it as it was during the Apollo lunar landing era. It will be cleaned and the consoles put into the proper locations. Spurious equipment will be removed. The wall-mounted P-tube station will remain. There will be appropriate lighting and the consoles will show lights and event indicators.

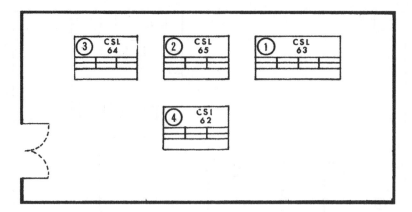

Figure 5.34 The Simulation Control Room. Drawing courtesy of Philco-Ford.

As currently planned, none of the SSRs are being restored, so this room could be illustrative of the others. Its consoles were among the first to be shipped to the restoration contractor, Cosmosphere/Space Works.

5.4.6   Visitor Viewing Room

The original Mercury Control Center included a room that overlooked the main control room to accommodate the many dignitaries and other visitors that wanted to see the state of affairs. People from all over the world visited the control center during missions. It epitomized the American way of conducting spaceflight in the open. It also didn't hurt to give the congressmen and senators funding spaceflight programs a front row seat for the epic events of the time. Astronauts' wives and children were often in the "VIP" room, as occasionally were members of visiting royalty. It was therefore decided that the new MCC-H should also feature such a viewing room.

When the MCC-H was designed, room #332 directly behind the MOCR was 56 feet long and 26 feet wide, and could accommodate 74 visitors in five tiered rows of theatre-style seats. On occasions there were so many visitors that some had to either stand or sit on the floor. They looked through a wall of glass down on the flight controllers and the large wall screen displays. Those occupying the front row could plug in headsets that listened in to the flight directors and other conference loops. The front row was usually reserved for astronaut families, or dignitaries. A standing desk was installed behind the top row. In the two corners were two communications booths and on the other corners two telephone booths each.

Decades after the Apollo 11 mission, modifications were made to the area to accommodate tours. These included extending the elevator to the third floor and improving access to the room, in particular ADA-compliant access. Modern flat-screen monitors were added, as were murals and photographs on the walls. It was now a popular tourist area.

The restoration will include:

- Cleaning and restoring all the upholstered seats.
- Removing non-Apollo photographs and signage.
- Modifying the lighting to reduce the ultraviolet radiation.
- Repairing the standing desk.
- Placing black rotary telephones in the booths.
- Cleaning and repairing the carpets.
- Configuring the room to how it was for the Apollo 15 mission in 1971.

As very large numbers of tourists will likely visit this area over time, it will be difficult to keep them from damaging the area, so some restrictions on them will be required, such as "Please don't put your drink down here," and "Please do not sit on this."

Figure 5.35 The Visitor Viewing Area in 2017. Photo courtesy of NASA.

5.5 COMPUTER FACILITIES

5.5.1 Real Time Computer Complex (RTCC)

By the early 1960s, having supported a wide range of projects that included Echo and Saturn, and the computing complex at the Goddard Space Flight Center that supported Project Mercury, IBM was deeply involved in the U.S. space program. In April 1962 they were awarded the Gemini Onboard Computer contract and, in October the Houston contract for the Ground Based Computing System that was later named the Real Time Computer Complex.

By January 1963 the first IBM 7094 to be used for software development was delivered to the temporary IBM facility on the Gulf Freeway in Houston. Later it received two other such machines. Eventually, some 600 IBM people worked on the project, two-thirds of them assigned to software development. The magnitude of the task was greatly underestimated both by IBM and NASA. Hardware needs increased along with the staff and NASA requirements.

Once Building 30 at the Manned Spacecraft Center became available the first three machines were moved from their temporary locations to room #112 on the first floor. Two more computers were subsequently added. The size and rating of the machines was also increased to model 7094-11s, with 65,000 words of main core storage and 524,000 words of additional core as a fast auxiliary memory. In this configuration, one machine was the Mission Operational Computer (MOC) and a second was the Dynamic Standby Computer (DSC). The third was known as the Simulation Operations Computer as before. One of the two new machines became the Ground System Simulation Computer. The other was a standby for future software development. The Ground System Simulator functioned like the tracking network, telemetry network, and related ground-based parts of mission control to facilitate the testing of other software.

The step from Mercury to Gemini entailed an order of magnitude increase in control center functionality. Mercury spacecraft telemetry consisted of less than 100 parameters in analog or binary format. All such telemetry was processed by analog ground stations and displayed using lights, analog meters, and strip chart recorders at the station which received it. There were no uplinked commands. In contrast, Gemini telemetry consisted of a digital downlink with several hundred measurements. Uplink commands and onboard recorded telemetry (replayed to the ground) were also included. Some ground stations were linked to Houston using land lines and others were connected by TTY to the RTCC.

When the MCC-H was designed, there were only three mission operational functional systems:

- Communications, Command and Telemetry Systems (CCATS).
- Display/Control System.
- Data Processing System/Real Time Computer Complex (RTCC).

The RTCC system performed the necessary operational, real time computation tasks, and also generated data in support of the system testing and flight controller training and simulation activity. Each of the five computing systems in the RTCC consisted of the following:

- Real Time Computer Subsystem.
- Computer Control Subsystem.
- Auxiliary Data Processing Subsystem.

Over the years, these systems were upgraded as the technology advanced. The following sections describe the system as it was introduced for the early Gemini missions. The major upgrades will be discussed later.

Real Time Computer Subsystem
This subsystem received the incoming data, stored and recorded it, and made the necessary computation and analyses, formatted it for display, and passed it to the users,

principally the flight controllers in the Mission Operations Control Rooms and Staff Support Rooms on the floors above. The first five IBM 7094-11s in the RTCC supported the Gemini missions and the first three Apollo-Saturn missions.

Figure 5.36 The RTCC for Gemini in 1966. Photo courtesy of NASA.

Specific components included:

- The Central Processing Unit (five IBM 7094s).
- Data Input/Output Multiplexer.
- Core Storage Unit (five IBM 2361s & two IBM 1460s).
- Data Channels.
- Direct Data Channel.
- Core Storage File.
- Data Communications.
- Computer Controller Multiplexer Unit.
- System Selector Unit (MOC, DSC and Off-line).

Basically, the system processed all the tracking and spacecraft telemetry data from the remote sites, processed all the commands from the control center to the spacecraft, and computed the current trajectory of the spacecraft from launch to orbit and all orbital maneuvering through to entry; probably the most complex of all the computer's functions. All the while, the RTCC was sending information in real time to all the displays and event lights used by the flight controllers on their consoles in the MOCRs, SSRs, and Simulation area.

The 7094s where state-of-art for their day. The software was mostly written in IBM Assembler Language, although the trajectory code was in FORTRAN. The computer was limited in processing speed, core storage, and tape drive capacity. The total memory of a system was 64K, with 36 bit "words." The system did not have any hard disks. Data were stored on tapes. The computer had 8 tape readers that were arranged in 2 banks. The tape readers/recorders copied up to 170,000 "characters" per second.

The heart of the system was a processor which ran at a frequency of 1 million instructions per second (MIP). This equated to about 500,000 logical operations, 250,000 additions, 100,000 multiplications, or 62,500 divisions per second. The data were displayed using black-and-green digital TV monitors and binary event indicators; the same units as the flight controllers used. During development and software testing, memory dumps were often taken and written out on high speed printers. Each computer occupied an area of 160 square feet of floor. Incoming tracking and telemetry data, and outgoing command data, were interfaced to the RTCC through the two UNIVAC 490s of the Communications, Command and Telemetry System (CCATS).

Computer Control Subsystem
This subsystem enabled the operators (listed below) to monitor the performance of a computer and manually control certain aspects of the mission programs and raw data selection. There were two complete control areas for this purpose with common consoles, displays, and manual entry devices.

The RTCC was staffed with five different positions (their call signs are given in parentheses) who manned consoles just like the MOCR flight controllers, but using with different displays:

- Radar data select (Select).
- Trajectory (Computer Dynamics).
- Command (Computer Command).
- Telemetry computations (Computer TM).
- Computer Supervisor (ComSup).

Each position in the RTCC was responsible for supporting specific MOCR flight controller disciplines. The ComSup was usually a NASA employee. Most other positions were manned by either an IBM employee or a NASA employee, but during critical mission phases such as launch and entry they were primarily NASA employees.

Auxiliary Data Processing Subsystem
This prepared computer input data, and recorded it on magnetic tapes for high speed input to the Real Time Computer Subsystem. Two identical groups had a central processing unit, a console printer, a card read/punch unit, printers, and magnetic tape units.

Upgrades

The Apollo Computers
After the final Gemini mission in November 1966 and the Apollo 1 accident in January 1967, MSC and IBM took advantage of the stand down to upgrade the RTCC. The 7094s were replaced by five IBM 360-75Js with the new IBM 360 operating system.

The major IBM hardware components of the system included:

- 2075 Processing Unit (750 nanosecond cycle time).
- Four 2365 Processor Storage units (1,048,576 bytes).
- 2361 Core Storage.
- 2860 Selector Channels.
- 2870 Multiplexor Chanel.
- 2150 Console with System Control Panel.

These were the computers that supported all of the manned Apollo missions. Again, the new software was a major step up. There were more mission phases, more vehicles capable of greatly increased telemetry downlinks and command uplinks, new trajectory algorithms, onboard navigation sources, more MOCR flight controllers, and the use of both the Space Tracking Data Network (STDN) and the Deep Space Networks (DSN).

Figure 5.37 The IBM 360 used in the RTCC during Apollo. Photo courtesy of NASA.

Post-Apollo
After Apollo, the RTCC was once again upgraded to the IBM 370 to support the Space Shuttle program, which was an even more demanding function. And after the loss of Challenger in 1986, NASA used the stand down to upgrade the RTCC to the IBM 3083s before the resumption of flights in 1988. But that configuration was short lived. In

September 1990, IBM announced its new family of Enterprise Systems Architecture (ESA)/390 operating systems. This was a new "connection architecture" with many functional enhancements.

But the computing world was moving away from large mainframe computers to distributed architectures. For the International Space Station (ISS) there would be new control rooms using networked UNIX workstations, with high resolution color displays and laser printers.

Over half a century has elapsed since the Houston MCC became operational, and in that time the "control center" world has drastically changed.

5.5.2   Auxiliary Computer Room

The Apollo Real Time Auxiliary Computing Facility (RTACF) was an extension of the facility available in the Gemini era. It was expanded to provide support for all aspects of flight control, with computer programs being developed to address both mission and mission-simulation support. Its scope was expanded to include prime mission support functions in addition to engineering evaluations. Hence it became a mandatory mission support facility, even though it was not in Mission Operations Wing M, being instead in the adjacent Administrative Wing A.

The facility functioned as a full scale mission support activity until after the Apollo 11 mission, then it gradually reverted to an off-line and on-call service. This downgrading reflected the increased capability and flexibility of the RTCC for the lunar landing missions. Once verified, these new capabilities eliminated the need for redundant computations. While the term "real time" was initially in its name, the RTACF was rarely used in real time, because the RTCC supported that function. It was, however, used to support the ongoing mission. The RT was dropped in the official manning list to just ACF. In part, this switch reflected the sense that the acronym RTACF was too similar to RTCC. Later it became ACR, which was Auxiliary Computing Room.

In addition to making mission-critical computations for unexpected trajectory computations, and engineering evaluations of trajectory problems that developed during the missions, the ACR was expanded to perform the following functions:

- Computations that required multiple ephemerides and trajectory planning beyond the time-limited capability and flexibility of the real time program.
- Computations using large and cumbersome programs with running times that were too slow to be feasible in real time computations.
- Computations to satisfy requirements that were established too late in the real time program planning cycle to be achieved by the real time system. On later missions, however, these computations were normally absorbed by the real time system.
- Computations performed to provide a real time testing ground for future RTCC programs. This offered a way to evaluate a computer program in a simulated real time environment before the program was implemented in the real time program.
- Computations to provide unlimited real time flexibility regarding the type of problem that could be resolved using the engineering analysis programs by the personnel who were involved in all phases of the mission planning.

The ACR had its own IBM 7094 (Mod 1) and its supporting panoply of tape drives, high speed printer, card reader, online printer and control console. It also had an IBM 1401 computer that was employed to load jobs onto tape and build the job stream for the 7094, read output tapes, and print user output. The facility was effectively an extension of Building 12 and its equipment was managed and operated by the Building 12 contractor, Lockheed. This separation meant that a Mission Planning and Analysis Division (MPAD) employee wishing to submit a job to the Building 12 computers had to carry card decks back and forth.

The responsibility for this kind of work fell within the purview of the Flight Analysis Branch of MPAD. Since they were already rather busy with real time functions, it was decided to form a separate small group composed of engineers who were familiar with most parts of mission planning and mission operations. This group had the responsibility of organizing and managing the overall ACR function and reporting to MPAD. It included NASA and contractor support.

To synchronize with the RTCC, and to respond to flight control demands for nominal and off-nominal mission situations the processing was both formalized and well organized. Its activities were different for the pre-launch planning of a mission and for subsequent simulation and real time control. During pre-launch planning, the assigned personnel were engaged primarily in deciding which of a list of requirements could be accommodated, in assembling the proper computer programs, verifying exercises, developing operating procedures, developing and coordinating work schedules, assigning other personnel to specific support, and developing the required data base. During the simulation and real time mission control activities, some personnel from this organization were supervisors and partial staff for the Flight Dynamics SSR and the ACR.

The room was staffed with three shifts and included the following positions:

- ACR Chief.
- Trajectory Consultants.
- Program Consultants.
- Computer Run Coordinators.
- Engineering Aides.

The major contribution made by the ACR to the Apollo program was probably building confidence in the RTCC program during preparations for and during the first lunar landing mission.

In addition, numerous specific computations that could be performed only by the ACR contributed significantly to the success of Apollo. The following were particularly noteworthy:

- Support for the Earth resources photography experiment (Apollo 9).
- Telescope pointing data (all lunar missions).
- Mass properties and entry aerodynamics for RTCC initialization (all missions).
- Launch pad abort impact points (all missions).
- Onboard navigation support (Apollo 8).
- Passive thermal control attitudes (all lunar missions).
- Pointing data for the 210-foot antenna at Goldstone, California (all lunar missions).

Figure 5.38 Lockheed contractors working with the ACR IBM 7094s during Gemini. Photo courtesy of NASA.

Figure 5.39 The back work area of the ACR. Facing away from the camera on the far left is Hector Garcia, to his right is Ted Turner, and in the foreground is Jerry Kahanek. Facing the camera on the right is Bob Regelbrugge, and to his right in the white shirt is Bill Reini. Photo courtesy of NASA.

Figure 5.40 Bill Sullivan checks a printout in the ACR. Photo courtesy of NASA.

- Lunar orbit insertion crew chart data (all lunar missions).
- Optimized translunar midcourse correction targeting (Apollo 8 and 10)
- Trans Earth midcourse correction (Apollo 8).
- Entry tracking ship positioning (all lunar missions).
- Solar flare data reduction (all lunar missions).
- Powered descent abort polynomial coefficients for the lunar module onboard computer (Apollo 11).
- Verification of all maneuvers performed (all missions).

See Appendix 2 for the Mission Manning Lists of those who worked in the ACR.

5.5.3   Communications, Command and Telemetry Support (CCATS)

Sometimes, you need a "middle man" to make things work. If ever there was a group in the middle of the action, it was the people in the CCATS, pronounced "Sea Cats." Anything that anyone in the MOCR wanted to do wouldn't happen without them. Every voice communication, telemetry signal, tracking data and spacecraft command went through the CCATS. And every signal that passed to each spacecraft, astronaut or experiment, whether in orbit or on the Moon, went through the CCATS. That included the biomedical, television, radar signals and systems information. The CCATS connected the Mission Control Center to the Space Tracking and Data Acquisition Network (STADAN), the Manned Space Flight Network (MSFN), the NASA Communications (NASCOM) Network, the Apollo Launch Data System (ALDS), and the Deep Space Network (DSN).

For Apollo, the outside world to which the CCATS was connected comprised 14 remote stations, each with two computers, as well as computers at other sites. It was a total of 39 Univac 642B computer systems (to replace the 1218s used in Gemini) to relay telemetry and command information between Houston and the Apollo spacecraft.

There were:

- Manned Space Flight Network (MSFN) stations (pronounced "Missfin")

  1. Guam.
  2. Kauai, Hawaii.
  3. Corpus Christi, Texas.
  4. Cape Kennedy, Florida.
  5. Grand Bahama Island.
  6. Bermuda.
  7. Antigua.
  8. Grand Canary Island.
  9. Ascension Island.
  10. Guaymas, Mexico.
  11. Merritt Island, Florida.

- Deep Space Network (DSN) stations referred to as "Wing Stations"

  12. Goldstone, California
  13. Madrid, Spain
  14. Canberra, Australia.

The CCATS was also connected to five Apollo Instrumentation Ships (AIS) that provided crucial tracking, telemetry, command, and voice communications. Three were insertion & injection ships: Vanguard, Redstone and Mercury. Two were reentry ships: Watertown and Huntsville. There were also eight EC-135A Apollo Range Instrumentation Aircraft (ARIA) for voice communications and telemetry recording.

Only two satellites were available for relaying data during Apollo, an Intelsat over the Atlantic and one over the mid-Pacific. Connected stations could relay to Goddard and Houston. The Tracking and Data Relay Satellite System (TDRSS) used nowadays was only a dream in the Apollo era.

The CCATS real time system decommutated and distributed incoming data into the RTCC, console displays, and assorted monitoring devices. It formatted and then issued command data to Goddard for distribution to the desired remote sites. In effect, the 39 remote site UNIVAC 642B computers talked to the three UNIVAC 494 computers in the CCATS, which talked to five IBM computers in the RTCC, to give the fight controllers in the MOCR what they needed to carry out mission operations.

For Apollo, the CCATS was located on the first floor of Building 30 adjacent to the RTCC. The CCATS consoles (with call signs in parentheses) included the following:

- Command Support Console. Three positions including:

  1. Real Time Command Controller (RTC).
  2. Command Load Controller (LOAD CONTROL).
  3. CCATS Command Controller (CCATS CMD).

- Telemetry Instrumentation Control Console. Two positions including:

  1. Telemetry Instrumentation Controller (TIC).
  2. CCATS Telemetry Controller (CCATS TM).

- Instrumentation Tracking Controller Console. Two positions including:

  1. Instrumentation Tracking Controller (TRAK).
  2. USB (Unified S-Band).

- Central Processor Control Console. Two positions including:

  1. Central Processor Controller (CPC).
  2. Central Processor Maintenance and Operations (M&O).

- Communications Controller Console.

Note that the first three areas listed above were part of the Instrument Support Team. The primary interface between the CCATS and the MOCR passed through the Network Controller and his Network Support Team.

The CCATS room operated on three or four shifts, depending on the mission, with each shift having about 20 people. There were also many peripherals in the room such as magnetic tape units, high speed teleprinters, card processors, buffer terminals, modems, and TTY terminals.

All digital data that was routed through CCATS was processed by one of the three UNIVAC 494s that were adjacent to the CCATS operators. Normally one computer served as the Mission Operations Computer (MOC) and another as the Dynamic Standby Computer (DSC). In the event of the MOC failing, an operator would manually switch over to the DSC. The third computer in this facility was used for checkout, debugging, and local operations.

5.6 OTHER SYSTEMS AND EQUIPMENT ROOMS

5.6.1 Display Projection Rooms

When someone walks into the Visitor Viewing Area, the first things they see are the large screen displays that dominate the far wall. As part of the restoration of the Apollo MCC, there will be an effort to preserve these iconic displays. Oddly enough, in the original conceptual design for the Mercury Control Center in the late 1950s (even before the first spaceflight) it was not clear that displays of this type would be useful. As it turned out they are very useful because they give the flight controllers an immediate visual representation of the status of the mission from different perspectives; a difficult thing to achieve without such displays or by some other means.

Having proved their worth, the group displays in the Mercury Control Center were periodically upgraded. When the control center was adapted for Gemini in 1964, the major changes were in the large screen group displays. These concepts were carried over to the design of the new MCC-H and expanded significantly to reflect advances in display technology and the requirements of flight controllers. The Mercury map was a mechanical

Figure 5.41   A UNIVAC 494. Photo courtesy of Sperry Rand.

wonder incorporating wire cables and servo motors. The side projections were not much more than projected vu-graphs. The MOCR displays in the new MCC-H were a combination of new technology and old technology.

When the MCC-H was designed, the flight controller requirements included many different types of displays. Almost all of these displays were controlled to some degree by the RTCC and selected by a flight controller as appropriate. The Display/Control System included dynamic and reference displays, and made use of plotting, television, and digital methods to present the data. For example, the dynamic displays would include real time telemetry from the spacecraft systems and crew biomedical data. Reference data included such information as mission rules, operational procedures, nominal sequences or historical performance data. The new system included ways of formatting such data for the flight controllers. Some of the displays would go directly to the flight controllers' consoles, while other data would go to the group displays. Both types will be preserved to some extent for the restored Apollo Mission Control Center, but not using the original technology.

One of the most visible tasks for the restoration people will be to recreate the large screen displays. The original technology was a Projection Plotter Display that comprised a set of seven projectors: one background, two spotting, and four scribers. The background projector displayed the world map from a square slide that was 1 inch square. The spotting projectors imposed symbols to represent the spacecraft or a target on the map, then relocated the symbols to match trajectory data supplied by the RTCC. The computer also controlled the scribing projectors that used diamond-tipped styli to scratch alphanumeric character or an X-Y plot through the metalized coating on a glass slide. A 2500 watt xenon lamp projected the slide onto the screen. There was also an Eidophor projection system which is now obsolete. This produced the images most people are familiar with in videos or photos of the control center. This complex optical technology will have to be replaced during the restoration. It was part of the Group Display Subsystem that provided varying degrees of information to three control areas: the MOCR, the Staff Support Rooms and the Recovery Control Room, as well as to the RTCC and the Mission Briefing and Observation Auditorium which was located in the Administration Wing.

There were four 10 x 10 foot projections screens and one 10 x 20 foot screen. They could each use either or both systems. The Eidophor used folded optics so that the optical throw distance required for quality projections could be achieved in a shorter space. Nevertheless, the rear projection room was very large (36 feet wide, 65 feet long, and 15 feet high) with its walls, ceilings and floors all painted black; no wonder it was known as the "Bat Cave." The Eidophor was replaced in the early 1980s with GE light-valve type projectors; they, in turn, were replaced in 1990.

In the late 1980s, during the Shuttle era, it was decided to further upgrade the display system. Although the existing system provided extremely sharp, bright functional displays, new missions were more demanding. Several studies finally set the requirements for the Request for Proposals. This said that a replacement for the center screen was not commercially available, but the side screens were. After further study, a request was issued in 1986 that specified the requirements for a large screen projector. Six qualified companies responded; two were direct laser projection products, two were oil-film projectors, and two made use of the polarizing light characteristics of crystals. The contract went to Hughes Aircraft Company for their HDP-6000B projector that used liquid crystal displays. After fabrication and testing, the new projectors were installed in both MOCRs during 1990. Each installation consisted of the projector, new screen, projector lift table, workstation/graphics processor, application software, maintenance monitors and maintenance documentation.

Although room #316 behind the large screen in the third floor MOCR that is being restored will not be visible to visitors, it is considered an extension of the MOCR and should be treated in the restoration to some degree. In fact, there is another area even behind the projection area (the "Bat Cave") that is even larger and was originally filled with display control terminal equipment. Other modern means of projection will be used to give the viewer a high fidelity representation of Apollo displays.

Because the display and control system served both MOCRs as well as other support rooms, the equipment was on all floors and in many rooms.

5.6.2   Equipment Areas

A building of the size and complexity of the MCC-H includes many rooms that are used for mechanical equipment, heating and air conditioning, testing, facility maintenance and telephone and voice communications and storage. There were also restrooms on either end of each floor. Some of the specific equipment on the first floor includes:

- Voice & Intra-Facility Communications Equipment Room #116.
- Teletype Message Center Room #119.

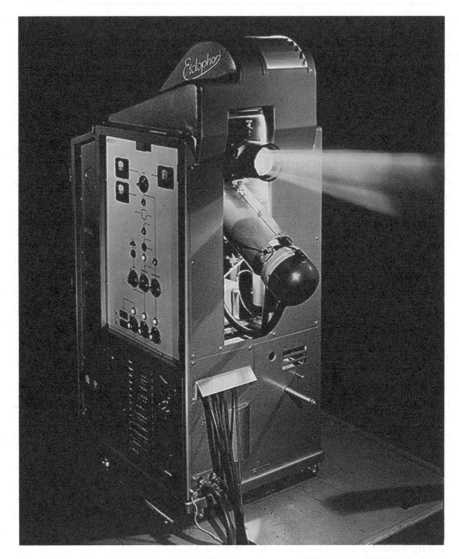

Figure 5.42 An early Eidophor, circa 1959. Photo courtesy of Philco and ETH-Bibliothek.

- Battery Room #122.
- Main Distribution Frame & Common Carrier Equipment Room #127.
- Telco Liaison & Maintenance Room #128.
- Tape Storage & Auxiliary Equipment Room #110.

In addition, there were various mechanical equipment rooms on the other two floors. There was the Emergency Power Building #48 of 10,500 square feet that contained all the standby equipment for power generation, air-conditioning, and interim and backup heating facilities for Building 30.

Figure 5.43 Projection plots of descent stage engineering data for Apollo 11. As Retrofire Officer (RETRO) Chuck Deiterich explains: On the flight path angle versus velocity plot "A," the "B" curves had time delay built in (for processing delays) and would allow the Lunar Module (LM) Ascent Propulsion Systems (APS) thrust to overcome a high descent rate. The Flight Dynamics Officer (FIDO) monitored the "A" and "C" plots. His main task was to compare the trajectory sources and advise the Flight Director of any deviations or differences. He would call an abort if line "B" was touched. He watched for onboard navigation errors. If an abort was called, the LM descent would be aborted. No aborts would be called after the crew took over with manual control. The Guidance Officer (GUIDO) monitored the "D" and "E" plots. His main task was to monitor the condition of the LM Primary Navigation Guidance System (PNGS) and Abort Guidance System (AGS) and note any differences, as well as how the programs were behaving; e.g. alarms. He also watched the onboard altitude and determined whether the landing radar should be incorporated into the onboard navigation solution. The RETRO task was to assure the onboard mass properties were correct before descent. During powered descent he monitored the guidance, and when the onboard computer telemetry indicated one minute to throttle down he would determine the elapsed time to throttle down and pass this to the crew. Photo courtesy of NASA.

6

The Missions

Project Mercury Project taught us that man can survive in space, he can perform well in weightlessness, and he can survive reentry. At the time, there were many who questioned whether this would be the case. Mercury also taught the aircraft industry how to produce a spacecraft. It taught NASA how to build a worldwide network to monitor the flights, and send information back and forth between the control center and the remote tracking sites. It taught the military how to modify missiles to transport humans, a role that the designers had never imagined. (The word "missile" was changed to "launch vehicle," which sounds somewhat safer). It also taught flight controllers how to support astronauts in space by being well trained in engineering and flight operations, and to build a control center for that purpose. That first NASA control center, the Mercury Control Center, supported Project Mercury from start to finish with very little in the way of revisions to its hardware systems and its operational methodology and procedures; it was well-conceived and designed for the missions being performed at that time.

The dream of Apollo focused our attention on the challenge set by President John F. Kennedy in a speech on May 25, 1961: "I believe that this nation should commit itself to achieving the goal before this decade is out of landing a man on the Moon and returning him safely to the Earth." That was even before we ever achieved orbit! The goal was there but not the details. A lot of basic capabilities would have to be proven before we could attempt to go to the Moon, land there, and return safely to Earth. That intermediate effort was promptly called Gemini, and it got underway even before Project Mercury was completed in 1963. There were people in the Space Task Group at the Langley Research Center as early as 1959 discussing the concepts; primarily based upon Langley's in depth research into aerodynamics, thermodynamics, and aircraft design. But first things first… Gemini had to prove more than that man could survive and perform useful work in space; it had to prove that NASA could perform the basic concepts that were required for lunar flights.

Many books have been written about Apollo. Suffice it to say that there were many successes, from the early unmanned Apollo tests through to the first lunar landing. Unfortunately, the Apollo 1 crew of Gus Grissom, Ed White and Roger Chaffee gave their lives for the program. But their sacrifice resulted in a better spacecraft that served the

© Springer International Publishing AG, part of Springer Nature 2018
M. von Ehrenfried, *Apollo Mission Control*, Springer Praxis Books,
https://doi.org/10.1007/978-3-319-76684-3_6

program well. It made possible the accomplishment of Kennedy's challenge. Had Grissom, White, and Chaffee died in space, or on the Moon, then that would have been the end of the U.S. manned space program; at least for decades. It is Apollo that best exemplifies what humans can do because "if you can put a man to the Moon, you can do anything."

Yes, there were follow-on missions, such as the Skylab and the joint mission with the U.S.S.R., and later the Space Shuttle, but the focus of this book is on the Apollo Mission Control Center and the missions that were flown from the iconic facility that is now a National Historic Landmark.

Here are the missions during the period 1964–1992, with the emphasis on the people who served in the MOCR 2 on the third floor.

6.1 GEMINI

Among those relocating from Langley, Virginia, to the new Manned Spacecraft Center in Houston in 1962 were those associated with the new Gemini Program Office. They worked with the designers of the Mercury capsule to specify a new two-man spacecraft capable of perfecting the flight operational goals that would be required for Apollo. Absolutely fundamental was the ability to maneuver the spacecraft, as a step towards demonstrating the ability to execute rendezvous in orbit.

As soon as the flight control teams had finished supporting Mercury, they too documented what they required in a new, more capable control center for future space programs. Much had been learned operationally during Mercury and these lessons had to be included in the design. The mission designers determined what Gemini would require to prove before a more complete design reference mission for Apollo could be firmly determined.

What Gemini had to prove was:

- Two spacecraft must be able to precisely maneuver in space for several reasons. They must change their orbits in order to rendezvous, and then maneuver in close proximity to dock together. In addition, they must be able to maneuver for precise reentry and landing.
- Astronauts must be able to work outside the safety of their spacecraft, in the vacuum of space, to carry out tasks in both Earth and lunar orbit and also on the lunar surface.
- It must be possible to operate in the space environment for 14 days, the equivalent of a round trip to the Moon with a visit to the lunar surface.

These operational requirements also affected the hardware and software that had to be built into the various vehicles and equipment.

These included:

- A new and more capable launch vehicle, the Titan II, developed as an ICBM, had to be man rated. It was called the Gemini Launch Vehicle (GLV).
- A target vehicle. The Agena was needed for rendezvous. When mated to the Atlas launch vehicle it was the Atlas-Agena Target Vehicle (GATV).

- The new two-man spacecraft had to be designed to achieve the necessary maneuvering capability and to have an onboard computer to calculate the orbital rendezvous.
- The Manned Space Flight Network had to be modified to interface with a new and more sophisticated spacecraft that was capable of both variable and higher orbits.
- The new Mission Control Center in Houston had to be more flexible, and have the improved computer, communications and command capabilities required to support the more complex missions of the new programs.

Project Mercury had merely established that man can survive weightlessness and the physical stresses of returning to Earth, so these requirements asked a lot of the Gemini program.

The following sections describe the Gemini missions in summary. These were the first missions to be flown from the new MCC-H. It was a learning experience for the people that trained the astronauts, plus the flight controllers and the many support teams and computer operators. It was essential preparation for the Apollo missions that would follow in later years.

All of the missions listed below, with three exceptions, were controlled out of MOCR 2. The exceptions were Gemini 1, 2 and 3, which were controlled by KSC MCC. Gemini 2 and 3 were monitored in Houston as the center capabilities came online.

6.1.1   Unmanned Flights

Gemini 1
Launched on April 8, 1964, this mission was simply to demonstrate and qualify the Gemini Titan Launch Vehicle (GLV) from launch through to orbit insertion. This included the malfunction detection system, guidance system, and structural integrity of the launch vehicle and the spacecraft. The Mercury Control Center, by then called the Mission Control Center, was manned and the flight dynamics controllers and systems were evaluated. This test also checked out the modified Manned Space Flight Network. MCC-H was not yet operational.

Gemini 2
Launched on January 19, 1965, this was a suborbital mission to demonstrate the ability of the heat shield to survive the maximum heating rate, and to determine the overall performance of the spacecraft systems. It also checked out the backup guidance steering signals. This flight was controlled by the KSC MCC but when this lost power shortly after launch the monitoring was shifted to a tracking ship. MCC-H passively monitored the short flight.

6.1.2   Manned Flights

Gemini 3
Launched on March 23, 1965, this "shakedown" mission was by astronauts Gus Grissom and John Young. It qualified the launch vehicle and spacecraft for later flights. They demonstrated the ability of the spacecraft to change its orbit. They returned to Earth after the planned three orbits, demonstrating a backup retrofire procedure. It also evaluated the new

Gemini tracking network. It was controlled from KSC MCC, but monitored and backed up by the new third floor MOCR 2. (About a month or so before the launch, Gus invited me to the launch pad to see "Molly Brown" on top of the Titan II launch vehicle. I had never been that close to a rocket or a spacecraft before. I thought to myself, "Boy, there is a lot to go wrong." At the time I was heavily into Mission Rules.)

Figure 6.1 Herbert Humphrey comes into the KSC MCC to congratulate the team. L-R: Richard Sutton, Larry Armstrong, and yours truly. Photo courtesy of NASA.

Gemini 4

Launched on June 3, 1965, astronauts Jim McDivitt and Ed White demonstrated the long duration capability of four days (66 orbits), longer than all the Mercury flights combined. White made the first American EVA, lasting 21 minutes. The spacecraft made in-plane and out-of-plane maneuvers and also demonstrated the capability of using the propulsion system as backup to retrorockets. A total of 11 experiments were conducted. This was the first Gemini mission to be controlled out of the MCC MOCR 2. All subsequent such flights would be controlled from there. (A few weeks before launch, I was told to go to a classified meeting with Ed White, Flight Crew Support guy John O'Neil, General Carroll Bolender from NASA Headquarters, and several others. This was my first meeting with a guard outside the door. The subject was Mission Rules for an EVA which was not yet decided or released to the general flight control team. This was also my first flight as Assistant Flight Director, and I stood next to Chris Kraft for a very exciting 21 minutes as White ventured out of the spacecraft.)

Figure 6.2 Pat White and Pat McDivitt in the MOCR congratulate their husbands during the Gemini 4 mission. Photo courtesy of NASA.

Gemini 5

Launched on August 21, 1965, Gordon Cooper and Pete Conrad demonstrated the ability of the spacecraft and crew to spend a week (120 orbits) in space. This set a world record. The mission also demonstrated rendezvous guidance and navigation by rendezvousing with a point in space. They evaluated the rendezvous radar with a transponder at the Cape. There were problems with the innovative fuel cells that provided electrical power, but the planned duration was achieved. A total of 16 of 17 planned experiments were accomplished.

Figure 6.3 John Aaron discussing the failed oxygen tank heater during the Gemini 5 mission with Flight Director Chris Kraft. Photo courtesy of NASA.

Gemini 6

In attempting a launch on October 25, 1965, astronauts Walter Schirra and Tom Stafford experienced the Titan II engine shutdown while still on the pad. By the Mission Rules this required an abort with the crew ejecting from the spacecraft, but Schirra had sensed no movement of the launch vehicle and so didn't trigger the ejection. This action certainly saved the rocket and spacecraft. No one really wanted to use the ejection system because the spacecraft was pressurized at 100 per cent oxygen, and it was no place to light up a rocket! It was very quiet in the MOCR because in that situation there was nothing the flight control team could do. After a harrowing wait of around 40 minutes to safe the rocket, the pad team retrieved the astronauts.

They launched on December 15, 1965, as the Gemini 6-A mission, on which they demonstrated the rendezvous assignment using Gemini 7 as a target. While in proximity they demonstrated station keeping. The two spacecraft came within one foot of each other. Various systems tests and experiments were conducted.

Figure 6.4 Three Flight Directors after Gemini 5 recovery. Chris Kraft, Gene Kranz, and John Hodge. Photo courtesy of NASA.

Gemini 7
Launched on December 4, 1965, ahead of Gemini 6-A, astronauts Frank Borman and Jim Lovell demonstrated a 14 day mission. Gemini 7 served as a rendezvous target for Gemini 6-A. They evaluated a lightweight pressure suit and performed 20 experiments and guidance during reentry.

Gemini 8
Launched on March 16, 1966, Neil Armstrong and David Scott followed up their rendezvous with the first docking with the Agena Target Vehicle. An unexpected attitude excursion prompted them to undock, at which point it was discovered the problem lie in the Gemini spacecraft. With the spacecraft tumbling at a high rate, Armstrong had to use the Reentry Control System to regain stability. A Mission Rule required that the mission be cut short, and reentry was performed after only six orbits of a planned three day mission.

Figure 6.5 The flight controller team monitors the Gemini 6 launch scrub which followed the failed Atlas Agena launch. Photo courtesy of NASA.

Figure 6.6 Gene Kranz, George Low and yours truly during Gemini 7. Photo courtesy of NASA.

Figure 6.7 Flight Director John Hodge during Gemini 8. Photo courtesy of NASA.

Gemini 9-A

After the prime crew of Elliot See and Charles Bassett were killed while flying a T-38, their backups Tom Stafford and Eugene Cernan launched on June 3, 1966. Their Agena Target Vehicle already having been lost, the mission was provided with a backup Augmented Target Docking Adapter. They rendezvoused with this vehicle, but were unable to dock because the aerodynamic shroud had failed to jettison properly. Stafford called it an "Angry Alligator." Cernan performed an EVA, but after over two hours of struggling with suit problems, and having no hand or foot restraints, he was not able to test the Astronaut Maneuvering Unit (AMU) supplied by the Air Force. They conducted other experiments and then reentered after 45 orbits.

Gemini 10

Launched on July 18, 1966, John Young and Michael Collins rendezvoused and docked with the Agena Target Vehicle and executed maneuvers to change their altitude. They then rendezvoused with the Agena from Gemini 8 but didn't dock. Collins performed two EVAs; one to retrieve a Micrometeorite Collector on the side of the Gemini 8 Agena. After performing other experiments they reentered on the 43rd orbit.

Figure 6.8 Elliot See and Charles Bassett were the original prime crew for Gemini 9. Photo courtesy of NASA.

Figure 6.9 Managers and Flight Directors during Gemini 10. L-R: Mission Director William C. Schneider, Flight Director Glynn Lunney, Flight Director Christopher C. Kraft Jr., and Gemini Program Office Manager Charles W. Mathews. Photo courtesy of NASA.

Gemini 11

Launched on September 12, 1966, Pete Conrad and Dick Gordon made a first-orbit rendezvous and docking with an Agena Target Vehicle. This simulated the LM rendezvousing with the lunar orbiting CSM. Using the Agena's engine, they attained an apogee of 850 miles, thereby setting a record for manned Earth orbit. Gordon performed two EVAs. After undertaking 11 of 12 planned experiments, they reentered after 44 orbits.

Figure 6.10 Gathering around the Flight Director's console during Gemini 11. Seated L-R: Cliff Charlesworth, Chris Kraft, Chuck Mathews, James Bates, and Dick Sutton. Standing L-R: Deke Slayton, Bill Anders, and John Young. Photo courtesy of NASA.

Gemini 12

Launched on November 11, 1966, Jim Lovell and Buzz Aldrin wrapped up the Gemini program. During three EVAs, Aldrin demonstrated that he could work successfully on tasks that required manual dexterity. After conducting many of the planned experiments they reentered after 59 orbits.

6.2 APOLLO

Apollo had the well-defined goal of landing a man on the Moon and returning him safely to Earth. To do so before the decade was out was just icing on the cake, but a driving force none the less. Apollo required far more hardware and software and operational capabilities than one could have ever imagined at the beginning of the space program in 1958. In hindsight, it is evident that some people were working on Apollo concepts as early as 1959.

Figure 6.11 Astronauts and flight controllers watch the Gemini 12 recovery. Photo courtesy of NASA.

On April 1 of that year NASA Headquarters invited representatives of its various field centers to serve on a Research Steering Committee for Manned Space Flight. This was led by Harry Goett, an engineering manager at the Ames Research Center who became Director of the new Goddard Space Flight Center in September. Goett and nine colleagues described the status of work and planning toward man-in-space at their respective organizations. This was two years before Alan Shepard made the first suborbital flight and Kennedy followed up by issuing his challenge!

It was evident that NASA leaders intended to aim high. The Goett Committee established its consensus on the priority of NASA objectives. They discussed the technical problems and the proposed solutions. The members heartily endorsed a lunar landing as NASA's major long range manned spaceflight goal. But, at that time the technology for launch vehicles was not up to the challenge of lifting the amount of mass that would be needed for a lunar mission. This capability would also be characterized by the rendezvous method chosen. So began the analysis of direct ascent, and rendezvous in Earth orbit versus rendezvous in lunar orbit. In the meantime, Wernher von Braun's group in Huntsville, Alabama, was already investigating some rocket concepts and the Saturn project was announced by the DOD in February 1959. Irrespective of the chosen path to the Moon, a family of rockets would be built.

Over the next two years, as Project Mercury forged ahead, industry, academia and NASA studies continued to work the lunar mission problems. By the end of 1962 the mission was better defined, and the decision for lunar orbit rendezvous determined the choices of launch vehicles. So contractors were selected to begin work on the detailed designs. Apollo was well underway.

Of course the history of Apollo is well recorded in tomes which fill libraries, and is beyond the scope of this book.

Almost all of the missions listed below were controlled from MOCR 2. The missions controlled from MOCR 1 will be pointed out. Only the summaries of missions and their primary objectives are included herein.

6.2.1   Unmanned Flights

Apollo-Saturn 201
Launched on February 26, 1966, this mission gained flight information on the structural integrity and compatibility of the launch vehicle and spacecraft, and confirmed launch loads.

Particular attention was paid to the separation events:

- Separation of the S-IB stage from the S-IVB, instrument unit (IU) and spacecraft.
- Jettisoning of the Launch Escape Systems (LES) and boost protective cover from the Command & Service Module (CSM).
- Separation of the CSM from the S-IVB.
- Separation of the Spacecraft LM adapter (SLA) from the IU.
- Jettisoning of the Service Module (SM) by the Command Module (CM).

It was to obtain flight operation information and performance of the following subsystems:

- Evaluate launch vehicle propulsion, guidance and control, and electrical systems.
- Evaluate CM heatshield, service propulsion system (SPS), environmental control system (ECS), communications (partial), the CM reaction control system (RCS), SM RCS, stabilization control system (SCS), Earth landing system (ELS), electrical power system (EPS) partial.
- Evaluate performance of the space vehicle emergency detection system (EDS) in an open-loop configuration.
- Demonstrate the mission support facilities and operations required for launch, mission conduct, and CM recovery.
- Recover the CM.

All of the primary objectives were achieved. This flight was controlled from MOCR 1.

Apollo-Saturn 203
Launched ahead of AS-202 on July 5, 1966, this evaluated the performance of the S-IVB/IU stage under orbital conditions in order to obtain flight information on:

- Venting and chill-down systems.
- Fluid dynamics and heat transfer to propellant tanks.
- Attitude and thermal control systems.
- Launch vehicle guidance.
- Checkout in orbit.

All of the primary objectives were achieved. This flight was controlled from MOCR 1.

Apollo-Saturn 202
Launched on August 25, 1966, it evaluated the CM heatshield at a high heating load and obtained further launch vehicle and spacecraft information on:

- Structural integrity and compatibility.
- Flight loads.
- Stage separation.
- Subsystem operations.
- Emergency detection system operation.

All of the primary objectives were achieved. This flight was controlled from MOCR 1.

Apollo 1 AS-204
Scheduled for launch on February 21, 1967, this was to have been the inaugural manned flight of the Apollo series, but a cabin fire during a "Plugs Out" launch rehearsal on January 27 killed all three crew members: Command Pilot Virgil I. "Gus" Grissom, Senior Pilot Ed White and Pilot Roger B. Chaffee. The tragedy was heard by the MOCR 1 team that was on-console to monitor the test; I was on the Guidance Officer's console. On April 24 the name "Apollo 1" was retired by NASA in commemoration of the crew. The events of that day have been written about many times. Here we just honor them for their sacrifice. See Appendix 8 for more details.

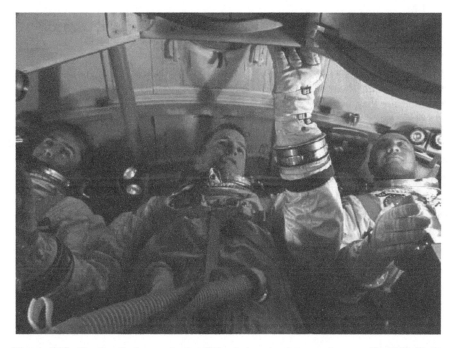

Figure 6.12 The Apollo 1 crew in the CM mission simulator on January 19, 1967. Photo courtesy of NASA.

Apollo 4 AS-501 (Note the nomenclature change)

Launched on November 9, 1967, this was the first "all-up" launch of the Saturn V launch vehicle and it evaluated the following capabilities:

- Verify pre-launch and launch support equipment compatibility with launch vehicle and spacecraft systems.
- Demonstrate the structural and thermal capabilities of all the stages of the Saturn: S-IC, S-II and S-IVB throughout the flight sequence and envelope, and establish their readiness for manned flight.
- Demonstrate the launch vehicle sequencing system.
- Evaluate the performance of the emergency detection system in an open-loop configuration.
- Demonstrate the mission support capability that was required for launch and mission operations to high post-injection altitudes.

All of the Saturn V primary objectives were achieved. The following Apollo Command & Service Module (CSM) capabilities were evaluated in the Saturn V launch environment:

- Evaluate the performance of the CSM emergency detection subsystem (EDS) in the open-loop configuration.
- Demonstrate the CSM/SLA/LTA/Saturn V structural compatibility and determine the spacecraft loads.
- Determine the force, acoustic, vibration and thermal environment of the simulated LM.
- Evaluate the thermal and structural performance of the Block II thermal protection system of the CM in a simulated lunar return, high heat load reentry.
- Demonstrate overall SPS performance.
- Verify the performance of the SM/RCS and CM/RCS systems.
- Demonstrate the performance of CSM/MSFN S-Band communications.
- Measure the integrated skin and depth radiation dose within the CM.

All of the primary objectives were achieved. This was the first Apollo mission to be controlled out of MOCR 2, namely the Apollo Mission Control Center now designated a National Historic Landmark.

Apollo 5 AS-204/LM-1

Launched on January 22, 1968, it was the first unmanned test of the Apollo Lunar Module (LM), and was to verify the operation of the following LM subsystems:

- Verify the ascent propulsion system and the descent propulsion system, the latter being the first throttleable rocket engine to be fired in space.
- Simulation a landing abort by the "fire in the hole" method.
- Evaluate LM staging.
- Evaluate the S-IVB/IU orbital performance.

All of the primary objectives were achieved. This flight was controlled from MOCR 1.

Apollo 6 AS-502

Launched on April 4, 1968, this flight was to further demonstrate the structural and thermal integrity and compatibility of the launch vehicle and spacecraft and verify the launch loads and dynamic characteristics (partially accomplished), as follows:

- Demonstrate the separation of the S-II stage from the S-IC stage and then the separation of the S-IVB stage from the S-II stage.
- Evaluate propulsion (including S-IVB restart), guidance and control and the electrical system.
- Evaluate the space vehicle EDS in a closed-loop configuration.
- Demonstrate mission support facilities and operations needed for launch, mission conduct, and CM recovery.

Most of the primary objectives were achieved, but some only partially. This flight was controlled from MOCR 2.

6.2.2 Manned Orbital Missions

Apollo 7 AS-205

Launched on October 11, 1968, this was the first manned mission in the wake of the Apollo 1 fire. During eleven days in Earth orbit it:

- Evaluated the redesigned Block II CSM.
- Practiced LM rendezvous with an S-IVB spacecraft adapter target.
- Evaluated the service propulsion system (SPS) with eight firings.
- Demonstrated CSM/crew performance.
- Evaluated mission support facilities performance.
- Conducted NASA's first live TV broadcast.

All of the test objectives were achieved. This was the only manned Apollo mission to be flown out of MOCR 1. (I was the Mission Staff Engineer in the Flight Director's SSR responsible for the flight test objectives.)

Figure 6.13 The Apollo 7 Flight Control Team. Photo courtesy of NASA.

6.2.3   Lunar Missions

Apollo 8 AS-503

Launched on December 21, 1968, this demonstrated the performance of crew, space vehicle, and mission support facilities including:

- First crew to be launched on a Saturn V.
- First crew to leave Earth and fly toward the Moon.
- Conducted the first Translunar Injection (TLI) burn.
- Conducted CSM navigation, communications, and midcourse corrections.
- Demonstrated primary and backup lunar orbit rendezvous (LOR).
- Assessed CSM consumables and passive thermal control.

All of the primary objectives were achieved. This flight and all later manned Apollo missions were controlled from MOCR 2. (I was in the Flight Director's SSR when the crew gave their readings from Genesis. It was a very memorable experience.)

Figure 6.14   MOCR 2 during Apollo 8. Photo courtesy of NASA.

Apollo 9 AS-504

Launched on March 3, 1969, this demonstrated crew, space vehicle, and mission support facilities performance during a manned Saturn V mission with CSM and LM including:

- LM/crew performance
- Performance of nominal and selected backup LOR mission activities, in particular:

    1. Transposition, docking, LM withdrawal.
    2. Intravehicular crew transfer.

3. Extravehicular capability.
4. SPS and DPS burns.
5. LM-active rendezvous and docking.
6. CSM/LM consumables assessment.

All of the primary objectives were achieved. This flight was controlled from MOCR 2.

Figure 6.15 MOCR 2 during Apollo 9. Photo courtesy of NASA.

Apollo 10 AS-505

Launched on May 18, 1969, this was a "full dress rehearsal" for the first lunar landing, and included:

- Testing all of the components and procedures, just short of initiating the powered descent to land on the Moon.
- Flying the LM down to an altitude of 8.4 nautical miles above the lunar surface, at the point at which powered descent would start on the actual landing mission.
- Demonstrating the performance of crew, space vehicle, and mission support facilities during a manned lunar mission with CSM and LM.
- Evaluating LM performance in the cislunar and lunar environment.
- Providing the flight controllers in MCC and the extensive tracking and control network with a high fidelity rehearsal for the procedures for a lunar landing.

All of the primary objectives were achieved. This flight was controlled from MOCR 2.

Figure 6.16 Apollo 10 Flight Directors. Photo courtesy of NASA.

Apollo 11 AS-506

Launched on July 16, 1969, this performed the first manned lunar landing and return. The detailed objectives and experiments included:

- Landing on the open plain of the Sea of Tranquility.
- Egressing from the LM and descending the ladder to the lunar surface, performing lunar surface EVA activities, and ingressing into the LM to prepare to lift off from the lunar surface.
- Performing lunar surface operations with the EMU/PLSS (spacesuit and backpack supplies).
- Obtaining data on the effects of DPS and RCS plume impingement on the LM, and data on the performance of the landing gear and descent engine skirt after touchdown.
- Obtaining data on the lunar surface characteristics from the effects of the LM landing.
- Collecting a contingency lunar sample.
- Collecting bulk lunar samples.
- Determining the position of the LM on the lunar surface.
- Obtaining data on the effects of illumination and contrast conditions on crew visual perception.
- Demonstrating the procedures and hardware developed to prevent back-contamination of the Earth's biosphere.
- Deploying the Early Apollo Scientific Experiments Package (EASEP) and conducting experiments.

- Providing television coverage during the lunar stay period .
- Obtaining photographic coverage during the lunar stay period.
- Lifting off from the lunar surface, rendezvousing in lunar orbit and then returning safely to Earth.

All of the primary objectives were achieved. This flight was controlled from MOCR 2. Figures 6.17 to 6.23 show some of the many flight controllers who supported Apollo 11.

Figure 6.17 RETRO Chuck Deiterich during Apollo 11. Others include Joe Engle, Dick Gordon, and Bob Parker. Photo courtesy of NASA.

Apollo 12 AS-507

Struck by lightning as it launched on November 14, 1969, this mission lost the CSM fuel cell power, instruments and telemetry. The crew recovered with help from the MCC. They went on to perform a precise lunar landing within walking distance of Surveyor 3, an unmanned craft that had touched down there in 1967.

The following lunar experiments were conducted:

- Landing on the cratered plain of the Ocean of Storms.
- Deployment and activation of the Apollo Lunar Surface Experiments Package (ALSEP) with six experiments.

Figure 6.18 Guidance Officer Steve Bales. Photo courtesy of NASA.

Figure 6.19 Flight Dynamics Officer Jerry Bostick. Photo courtesy of NASA.

Figure 6.20 Guidance Officer Granville Paules (nearest camera), Will Fenner (seated to Paules' left), Steve Bales standing, Bill Stovall (seated) and John Llewellyn standing at the end of the line. The bald man behind Bales' shoulder is Phil Shaffer. The picture was taken during the Apollo 11 mission. Photo courtesy of NASA.

- Lunar surface EVA operations, including a selenological survey and the collection of samples.
- Retrieved some parts of the Surveyor for analysis on Earth.
- Conducted two EVAs on the lunar surface.
- Obtained photographic coverage from orbit of candidate sites for future exploration.

Figure 6.21 EECOM John Aaron during Apollo 11. In the distance is GNC Buck Willoughby. Photo courtesy of NASA.

All of the principal objectives were achieved. This flight was controlled from MOCR 2.

Apollo 13 AS-508
Launched on April 11, 1970, it was to have flown to the Moon and landed at a preselected point in the hummocky Fra Mauro Formation in order to undertake selenological inspection, survey, and sampling, and also deploy and activate an Apollo Lunar Surface Experiments Package (ALSEP). Unfortunately, while on the way to the Moon an oxygen tank in the SM exploded, requiring the crew to transfer from the powerless CM into the LM in order to use that as a "lifeboat." Despite great hardship caused by limited power, loss of cabin heat, shortage of potable water and the need for makeshift repairs to the carbon dioxide removal system, the crew returned safely to Earth on April 17 after a flight of six days.

Figure 6.22 Flight Directors celebrate the successful return of Apollo 11. Photo courtesy of NASA.

Figure 6.23 Gene Kranz's team for Descent Orbit Insertion on the Apollo 11 mission. Photo courtesy of NASA.

While none of the mission objectives were met, this flight demonstrated the ability of the flight controllers in MOCR 2 and their support colleagues to come up with the work-arounds that were needed in order to return the crew to Earth. The crew demonstrated their ability to cope with the situation and to implement the necessary emergency procedures.

Figure 6.24   The MOCR two days before the explosion that crippled Apollo 13. Fred Haise is on the screen. Photo courtesy of NASA.

Apollo 14 AS-509

Launched on January 31, 1971, this mission:

- Landed on target in the hummocky Fra Mauro Formation.
- Performed selenological inspection, survey, and collection of samples.
- Deployed and activated an ALSEP.
- Evaluated the modifications to the oxygen system in the SM designed to prevent a recurrence of the Apollo 13 situation.
- Used the Modular Equipment Transporter (MET), a pull-cart for carrying equipment and samples during a traverse on the lunar surface.
- Conducted two EVAs on the lunar surface.
- Obtained 94 pounds of lunar material.
- Conducted CSM experiments, including photographing a target site that was assigned to Apollo 16.

All of the primary objectives were achieved. This flight was controlled from MOCR 2.

Figure 6.25 Astronauts and flight controllers gather around the console monitoring Apollo 13 on its way home. Standing on the immediate left is M. Peter Frank. Seated L-R: astronauts Alan B. Shepard Jr. and Edgar D. Michell, and Guidance Officer Raymond F. Teague. Standing behind them L-R: astronauts, Ronald Evans, Gene Cernan, Joe Engle, and Anthony England. Farther back are RETROS and FIDOS. Photo courtesy of NASA.

Figure 6.26 The crowded MOCR 2 for the Apollo 13 landing. Photo courtesy of NASA.

Apollo 15 AS-510
Launched on July 26, 1971, it:

- Landed on target on a plain set between mountains and a canyon in the Hadley-Apennine region.
- Performed selenological inspection, survey, and collection of samples.

- Deployed the first Lunar Roving Vehicle and used it to investigate the landing area.
- Deployed and activated an ALSEP .
- Conducted three EVAs on the lunar surface.
- Retrieved 170 pounds of lunar material.
- The SM deployed the PFS-1 subsatellite into lunar orbit.
- Conducted lunar orbit experiments and observations using the Scientific Instrument Module (SIM) of the SM to investigate the lunar surface and environment in great detail.
- During the voyage back to Earth, the CSM pilot made an EVA to retrieve several film cassettes from the SM.

All of the primary objectives were achieved. This flight was controlled from MOCR 2.

Apollo 16 AS-511
Launched on April 16, 1972, it:

- Landed on target in the Caley-Descartes region of the lunar highlands.
- Performed selenological inspection, survey, and collection of samples.
- Deployed a Lunar Roving Vehicle and used it to investigate the landing area.
- Deployed and activated an ALSEP.
- Conducted three EVAs on the lunar surface.
- Retrieved 211 pounds of lunar material.
- The SM deployed the PFS-2 subsatellite into lunar orbit.
- Conducted lunar orbit experiments and observations using the Scientific Instrument Module (SIM) of the SM to investigate the lunar surface and environment in great detail.
- During the voyage back to Earth, the CSM pilot made an EVA to retrieve several film cassettes from the SM.

All of the primary objectives were achieved. This flight was controlled from MOCR 2.

Apollo 17 AS-512
Launched on December 7, 1972, it:

- Landed on target in a valley between mountains in the Taurus-Littrow region.
- Performed selenological inspection, survey, and collection of samples.
- Deployed a Lunar Roving Vehicle and used it to investigate the landing area.
- Deployed and activated an ALSEP.
- Operated other experiments on the lunar surface, including the Surface Electrical Properties (SEP) experiment.
- Conducted three EVAs on the lunar surface.
- Conducted lunar orbit experiments and observations using the Scientific Instrument Module (SIM) of the SM to investigate the lunar surface and environment in great detail.
- During the voyage back to Earth, the CSM pilot made an EVA to retrieve several film cassettes from the SM.
- The mission broke several records: the longest Moon landing mission, longest total extravehicular activities, largest lunar sample, and longest time in lunar orbit.

- It ended the Apollo lunar program.

All of the primary objectives were achieved. This flight was controlled from MOCR 2.

In May 1973, with Apollo lunar exploration ended, MOCR 2 was deactivated to be reconfigured for the up-coming Space Shuttle program; the first such flight being in 1981. In the meantime, MOCR 1 was configured to support Skylab and the Apollo-Soyuz Test Project.

6.2.4 Apollo-Soyuz Test Project

Over a period of 7.5 hours, both the American Apollo CSM and the U.S.S.R.'s Soyuz were launched on July 15, 1975 into orbits in the same plane inclined at 51.8 degrees to the equator. After performing a rendezvous, the Apollo docked two days later using a one-off interface module. The Apollo was launched on a Saturn IB from the Kennedy LC-39B and the Soyuz by a Soyuz U rocket from Baikonur. This was the first joint U.S.-U.S.S.R. spaceflight, as a symbol of the policy of détente that the two superpowers were pursuing at the time. The plan called for docking Apollo CSM 111 with the Soyuz 7K-TM 19 spacecraft. The Apollo was surplus from the curtailed Apollo program, and the final one to fly. Tom Stafford, Vance Brand and Deke Slayton flew the Apollo. Alexey Leonov and Valeri Kubasov flew the Soyuz. This ceremoniously marked the end of the Space Race that began in 1957 with the launch of Sputnik.

This joint mission provided useful engineering experience for possible future US-Russian spaceflights. The prospect of joint ventures diminished with the end of détente, but after the fall of the Soviet Union the Shuttle-Mir program was a precursor to the development of the International Space Station.

Once in orbit, the Apollo spacecraft separated from the S-IVB and retrieved a module from the mount that would normally have carried a Lunar Module. This specially designed Androgynous Peripheral Docking System (APDS) was meant to enable the two spacecraft to dock and serve as a transfer compartment. While the two ships were docked, the three Americans and two Soviets conducted joint scientific experiments, exchanged flags and gifts, signed certificates, visited one another's vessels, ate together, and conversed in one another's languages. There were also docking and re-docking maneuvers in which the two vehicles reversed roles and the Soyuz was the "active" ship.

After 44 hours together, the two ships separated and maneuvered to enable the Apollo to create an artificial solar eclipse so that the crew of the Soyuz could take photographs of the solar corona. Another brief docking was made before the ships finally went their separate ways. The Soviets remained in space for five days, and the Americans for nine, during which time the Apollo crew also conducted Earth observation experiments.

6.3 SKYLAB

Devised in 1968, Skylab was the only part of the Apollo Applications Program to be implemented. Its overall objective was to develop science-based human space missions using hardware originally developed for the effort to land astronauts on the Moon. In effect it was a space station to enable astronauts to live in space for much longer than an

Apollo spacecraft would allow, and undertake a wide range of complex scientific experiments.

The Marshall Space Flight Center's Saturn team came up with the proposal to modify elements of a Saturn V as an affordable way to create a preliminary space station. Marshall developed and integrated most of the major Skylab components, including:

- The Orbital Workshop where the astronauts would live and work.
- The airlock module that would serve as a doorway to space for EVA.
- The multiple docking adapter that would enable an Apollo to dock and deliver people and equipment, with facilities for a second spacecraft to dock in order to perform a rescue.
- A mount with solar telescopes.
- A payload shroud to protect Skylab during launch.
- Many of the onboard experiments.
- Biomedical equipment that included a bicycle ergometer, a metabolic analyzer, a lower body negative pressure device, and the experimental support system.

Marshall provided the Saturn V that propelled the Skylab Orbital Workshop into orbit, and also the S-IB vehicles that launched the three crews.

All the Skylab missions were controlled from MOCR 1, which had not been used for an actual flight since Apollo 7 in 1968; nearly five years earlier. There were only slight modifications to accommodate the Skylab missions.

6.3.1   Unmanned Launch of Skylab

Skylab 1
On May 14, 1973, the final launch of a Saturn V propelled Skylab into orbit, but approximately 63 seconds into the ascent the meteoroid protection shield of the Orbital Workshop prematurely deployed. As it tore off in the airflow, it took one of the main solar panel arrays with it and jammed the other one in a manner that prevented to from deploying once in orbit. This left the station with a huge power deficit. As a result, the launch of Skylab 2 with the first crew was postponed. The NASA-industry team developed plans and hardware to enable astronauts to open the stuck solar array. In addition, a square thermal shield was fabricated to serve as a "parasol" to protect the exposed skin of the Orbital Workshop from the heat of the Sun. Don Arabian led the effort in the MER to conceive, develop, and test this parasol over a period of several days and nights, so that it could be delivered aboard Skylab 2.

6.3.2   Manned Missions

Skylab 2
An Apollo CSM was launched on May 25, 1973, by a Saturn IB as SL-2 with the first crew for Skylab. (This would be the final launch from LC-39A until SpaceX launched the CRS-1 Dragon to the ISS on February 19, 2017.) The initial mission made extensive repairs to the station. The crew Pete Conrad, Joseph Kerwin, and Paul Weitz deployed the folded "parasol" through a small instrument airlock from inside the station and opened it to form a sunshade. This lowered the temperature sufficiently to prevent overheating that

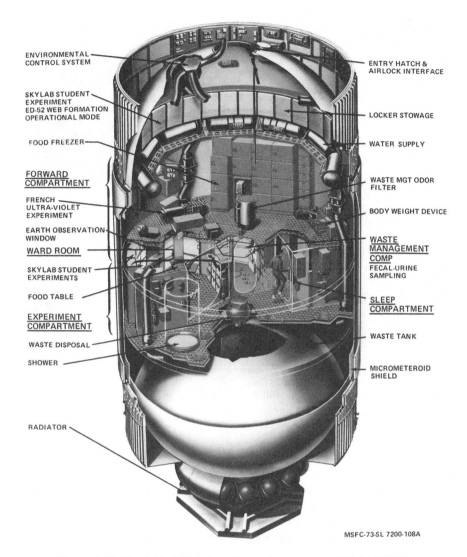

ENVIRONMENTAL
CONTROL SYSTEM

ENTRY HATCH &
AIRLOCK INTERFACE

SKYLAB STUDENT
EXPERIMENT
ED-52 WEB FORMATION
OPERATIONAL MODE

LOCKER STOWAGE

FOOD FREEZER

WATER SUPPLY

FORWARD
COMPARTMENT

WASTE MGT ODOR
FILTER

FRENCH
ULTRA-VIOLET
EXPERIMENT

BODY WEIGHT DEVICE

EARTH OBSERVATION
WINDOW

WARD ROOM

WASTE
MANAGEMENT
COMP
FECAL-URINE
SAMPLING

SKYLAB STUDENT
EXPERIMENTS

FOOD TABLE

SLEEP
COMPARTMENT

EXPERIMENT
COMPARTMENT

WASTE DISPOSAL

WASTE TANK

SHOWER

MICROMETEROID
SHIELD

RADIATOR

MSFC-73-SL 7200-108A

Figure 6.27 The Orbital Workshop. Illustration courtesy of NASA MSFC.

would have melted plastic insulation and released poisonous gases. Later the crew conducted further repairs on two EVAs. Upon finishing the assigned science experiments, the crew returned to Earth after the planned 28 days.

Skylab 3
SL-3 was launched on July 28, 1973, with Alan Bean, Owen Garriott and Jack Lousma. Early in the mission, Garriott and Lousma performed an EVA to erect a new twin-pole solar shield that provided better thermal control for the remainder of the Skylab missions. The second Skylab crew returned to Earth after spending 59 days in orbit.

Figure 6.28 The team crowds around Flight Director Don Puddy shortly after the launch of Skylab to discuss the solar array and heating problems. L-R: Frank van Rensselaer, Rod Loe, John Aaron, Dick Truly, Bill Harris, Bill Brady, Milt Windler, Neil Hutchinson, Phil Shaffer, and Don Puddy (seated). Photo courtesy of NASA.

Skylab 4
SL-4 was launched on November 16, 1973, with Gerald Carr, Edward Gibson and William Pogue. Although initially expected to be about as long as its predecessor, the mission was extended to 84 days. In addition to continuing the comprehensive research program the crew had an excellent view of Comet Kohoutek.

Skylab 5
SL-5 was to have been a short 20-day mission to conduct some more scientific experiments aboard Skylab and then use the Service Propulsion System of the docked spacecraft to boost it into a higher orbit, but it was cancelled. The crew members who trained for the flight were Vance Brand, William B. Lenoir and Don Lind.

Skylab Rescue
This mission was planned but never required. It would have been flown if it had become necessary to rescue a crew stranded aboard Skylab by a malfunction in their own Apollo spacecraft. Skylab had two docking ports. The Apollo capsule was adapted to carry five people. It would have launched with Vance Brand and Don Lind.

Figure 6.29 Skylab EGIL (EECOM) Sy Liebergot. Photo courtesy of NASA.

6.3.3 Skylab Reentry

Before SL-4 returned home on February 8, 1974, Skylab fired its thrusters for 3 minutes to add 6.8 miles (11 km) in height to its orbit. It was left in a 269 x 283 mi (433 by 455 km) orbit on departure, but that would continuously decay as a result of drag in the rarefied upper atmosphere. At that time, the date of reentry was estimated by NASA as nine years off. Although there were several plans to revisit Skylab, possibly by an early Space Shuttle mission, nothing came of the proposals.

In the hours leading up to reentry, NASA flight controllers in MOCR 1 made an attempt to adjust Skylab's trajectory and orientation, to minimize the risk of debris coming down in populated areas. NASA wanted it to fall into the Indian Ocean some 810 miles (1,300 km) south-southeast of Cape Town, South Africa. Skylab's atmospheric reentry began on July 11, 1979. It didn't burn up as fast as expected, with the result that debris fell southeast of Perth in Western Australia, leaving a trail of debris between the towns of Esperance and Rawlinna. (In fact, there is a Skylab museum in Esperance with many salvaged pieces.) Analysis of some of the debris indicated that the Skylab had disintegrated at an altitude of 10 mi (16 km); much lower than expected.

Between May 25, 1973 and February 8, 1974, three crews of three astronauts visited Skylab and undertook 270 scientific and technical studies which required 90 different

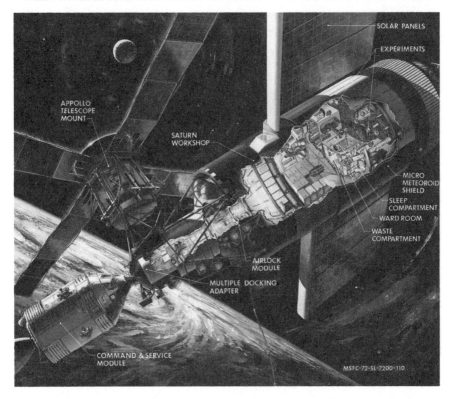

Figure 6.30 The overall Skylab configuration. Illustration courtesy of NASA MSFC.

pieces of experimental hardware. This research spanned the fields of physics, astronomy, and biological sciences. The three crews spent a total of 171 days in space, travelled more than 70 million miles, and logged over 41 hours of extravehicular activity. The final Skylab crew lived in space for 84 days, thereby establishing a new record for the longest human stay in space.

6.4   SPACE SHUTTLE

6.4.1   Enterprise Flights

The Approach and Landing Test (ALT) part of the Space Shuttle development involved 16 separate tests of Enterprise in three phases: "taxi-test," "Captive" (both Inactive and Active), and "Free Flight." These tests utilized the modified Boeing 747 known as the Shuttle Carrier Aircraft (SCA). While at the NASA Dryden Flight Research Center, Enterprise was evaluated for numerous ground and flight tests that were meant to validate aspects of the Shuttle program. The initial nine months of testing, from January 31 to October 26, 1977, included a maiden "flight" on February 18 in which the Shuttle remained atop the SCA to measure the structural loads and ground handling and braking characteristics of the mated system. Ground testing of all Orbiter subsystems were carried out to verify functionality prior to atmospheric flight.

Captive-Inactive Tests

The mated Enterprise/SCA combination was subjected to five "Captive-Inactive" flights. For these, the Enterprise was attached to the SCA for the duration of the flight and configured in an unmanned, powered-down state. This tested the flight and handling characteristics of the SCA while it was mated with Enterprise. The first three flights were flown with a tail cone affixed to the Orbiter's aft fuselage to reduce drag and turbulence when mated to the SCA. The purpose of these test flights was to measure the flight characteristics of the mated combination. After all such tests, Enterprise spent three and a half months on the ground so that its flight systems could be activated, tested, and prepared for the manned, Captive-Active Flights.

Captive-Active Flights

The "Captive-Active" flights were intended to determine the optimum profile required for Enterprise to separate from the SCA during the forthcoming free-flights. These were also intended to refine and test the Orbiter crew procedures and to ensure the operational readiness of its systems.

Enterprise remained mated with the SCA for these three flights, but it was powered and crewed by the following astronaut test pilots:

Enterprise "Captive-Active" flights

1	June 18, 1977	Fred Haise and Gordon Fullerton
2	June 28, 1977	Joe Engle and Richard Truly
3	July 26, 1977	Fred Haise and Gordon Fullerton

Free Flights

The final phase of Enterprise flight testing involved "Free" flights. The Orbiter was mated to the SCA and carried to a launch altitude, then separated by firing explosive bolts. The Orbiter would then to glide to a landing on the runways at Edwards AFB. The intention of these flights was to test the flight characteristics of the Orbiter itself, on an approach and landing profile typical of that intended for returning from orbit.

There were five of these flights between August and October 1977. The first three were conducted with the tail cone on Enterprise to reduce drag when atop the SCA. The final two flights had the tail cone removed, with the Orbiter in its operational configuration with dummy main engines and OMS pods. Enterprise used an air data probe on its nose for these flights. On the fifth and final gliding flight, pilot-induced oscillation problems were revealed, and these needed to be addressed before it was possible to attempt the first orbital mission. Enterprise was not capable of flying in space, so the inaugural launch would be performed by Columbia. These five flights were the only times Enterprise flew alone. The crews for these flights were:

Enterprise "Free" flights

1	August 12, 1977	Fred Haise and Gordon Fullerton
2	September 13, 1977	Joe Engle and Richard Truly
3	September 23,1977	Fred Haise and Gordon Fullerton
4	October 12, 1977	Joe Engle and Richard Truly
5	October 26, 1977	Fred Haise and Gordon Fullerton

Figure 6.31 The Enterprise ALT Flight Crew. L-R: Gordon Fullerton, Fred Haise, Joe Engle, and Dick Truly. Photo Courtesy of NASA.

During all this testing, the Boeing 747/SCA crew was:

- Fitzhugh L. Fulton, Jr. (captain).
- Thomas C. McMurty (co-pilot).
- Louis E. Guidry, Jr. (flight engineer).
- Victor W. Horton (flight engineer).

All these tests were monitored in MOCR 2, which had been modified for the ALT phase in August to December 1976 by the installation of the Approach and Landing Test Data System (ALTDS). After these flight tests, it was deactivated and the equipment removed to prepare MOCR 2 for the operational phase of the Shuttle.

Ferry Flights
Following the "Free" flights, Enterprise was prepared for "Ferry Flight" tests to verify the SCA/Orbiter configuration was viable for cross-country transportation flights. In March 1978, Enterprise arrived at the Marshall Space Flight Center in Alabama and taken to the Dynamic Structural Test Facility, where the complete stack of the Orbiter, External Tank (ET) and Solid Rocket Boosters (SRB) were subjected to vertical ground vibration tests to assess the structural responses to a number of scenarios. In April 1980, Enterprise was

transported to the Kennedy Space Center in Florida to fit check the facilities at LC-39A and the procedures that would be used in launching the Shuttle.

On December 12, 2011, ownership of Enterprise was officially transferred to the Intrepid Sea, Air & Space Museum in New York City. Enterprise was listed on the National Register of Historic Places on March 13, 2013.

Figure 6.32 Boeing Shuttle Carrier Aircraft flight and ground crews. The picture was taken prior to a pilot proficiency flight by NASA 747 SCA tail number 905 on March 29, 2012, and features eight of the nine pilots and flight engineers from JSC in Houston and Dryden in California who had flown on the two SCAs in recent years, along with the maintenance crew from Dryden. The ground crew in the front row are L-R: Randy Isaac, Eugene Smith, Robert Hackaday, Leroy Marsh and Rick Brewer, all of whom were employees of CSC Applied Technologies. Flight crew members in the back row are L-R: Henry Taylor, Larry LaRose, Frank Batteas, Bill Brockett, Arthur "Ace" Beall, Tim Sandon, Jeff Moultrie, and Bill Rieke. Pilot Bob Zimmermann is not present. Photo courtesy of NASA and Tony Landis.

6.4.2 Manned Orbital Flights

The manned orbital flights started on April 12, 1981 with Bob Crippen and John Young. It happened to be the 20th anniversary of Yuri Gagarin's spaceflight. The second floor MOCR 1 was now known as Flight Control Room (FCR) 1 and was used in support of STS-1 through STS-4 in July 1982, and then again in 1984. In 1973, with the lunar landing missions completed, MOCR-2 was deactivated and reconfigured to support first the Shuttle ALT and then the orbital flights, starting with STS-5 in November 1982. After supporting a total of 22 missions, seven of which were classified DOD flights, MOCR 2 (now FCR-2) was deactivated once again. Its last mission was STS-53 in December 1992. This duty reverted back to FCR-1 for the remainder of the program, including missions to the International Space Station (ISS). The final flight of the Shuttle program was STS-135 on July 8, 2011. In administrative terms, the program was formally ended on August 31, 2011.

Over the course of 135 missions, two Orbiters were destroyed with the loss of seven astronauts each time. Challenger was lost 73 seconds after it was launched on January 28, 1986 for STS-51L. It was being controlled out of MOCR 2. Then on February 1, 2003, Columbia was lost approximately 16 minutes before it was due to touch down in Florida to conclude the STS-107 mission. It was controlled from MOCR 1.

It would not be practical in a book of this scope to list all of the Space Shuttle manned missions, but some of the highlights were:

- The launch and servicing of the Hubble Space Telescope.
- Transporting supplies to space stations in Spacehab modules or Multi-Purpose Logistics Modules.
- Deploying and retrieving the Long Duration Exposure Facility.
- Deploying the Upper Atmosphere Research Satellite.
- Deploying the Compton Gamma Ray Observatory.
- Deploying the Earth Radiation Budget Satellite.
- Retrieving satellites for return to Earth.
- Mating the Docking Node to Mir.
- Deploying satellites with boosters that delivered the satellite to a higher Earth orbit, notably the first six TDRS satellites and the Chandra X-ray Observatory.
- A unique three-man EVA to capture an Intelsat satellite "by hand" and then fit it onto a new booster.
- Deploying two DSCS-III (Defense Satellite Communications System) communications satellites in one mission.
- Deploying a Defense Support Program satellite.
- Deploying interplanetary missions, including Magellan, Galileo, and Ulysses.
- Installing modules and other components on the International Space Station (ISS).

7

The People

As with many great adventures and achievements, there are the unseen and often unappreciated people that are down in the bowels of the system, working for the common goal of the team. In the 19th century, the ship captain would blow in a talk tube at the helm, and ask the crew to put on steam! In a building the size of Building 30 there were hundreds down in the bowels of the building "putting on more steam" in support of the mission. In modern parlance, they had their noses "to the grind stone." These were mostly cathode ray tubes and volumes of paper called Flight Plans, Console Handbooks, Operational Procedures, Mission Rules and such like. And, if one looked closely, their noses were, indeed, deep in to it. Over the course of half a century, many thousands worked in Building 30. The outside world saw only a very small percentage of them; perhaps no more than one per cent! The public had no idea of what was going on behind the views on their TV screens. They also knew nothing about the other people supporting the missions from other buildings and at other locations.

The public perception of NASA's space program was focused mostly on the astronauts. Views of the Mission Control Center usually featured them sitting at the CapCom's console. As the camera panned around the room, it would turn to the Flight Director, who was often standing. It rarely lingered at the consoles of the other flight controllers. The camera showed the "big screen," and the Public Affairs Officer would explain what was on that. The TV cameras never looked into the Staff Support Rooms, the other support rooms, or the communications, command, and computer rooms on the first floor or in nearby buildings.

During Project Mercury, there were occasional instances of live TV from the control room during a mission. The TV was mostly of a launch vehicle standing on its pad, in the hope of sighting the intrepid astronaut wearing his silver space suit.

Once Gemini came along with the new Houston Mission Control Center, the public became a little more interested in the program and more information was issued on what was going on in Mission Control. In the 1960s the evening news lasted only 15 to 30 minutes, and there were only three stations. There were not any 24 hour news broadcasts.

© Springer International Publishing AG, part of Springer Nature 2018
M. von Ehrenfried, *Apollo Mission Control*, Springer Praxis Books,
https://doi.org/10.1007/978-3-319-76684-3_7

Apollo 11 garnered plenty of interest. An estimated 600 million world-wide watched the coverage of the launch, lunar landing and moonwalk. (Who would have imagined we would see live TV coverage from the surface of the Moon!) But still, the camera was focused on the stars of the space program and not the people working in the other rooms.

By the time the follow-on programs came along, like the Space Shuttle, very few members of the public could name the members of a particular crew unless their participation was highly publicized, such as the first women astronaut or a teacher astronaut. Now, even fewer know who is aboard the International Space Station.

This book seeks to remember the people who worked in the National Historic Landmark now known as the Apollo Mission Control Center.

Appendix 2 contains the Mission Manning Lists with names and positions as researched by former Flight Director, Bill Reeves. There are other names given that are just as important but not necessarily on official lists. In some cases, the early pioneers of mission operations from over half a century ago are also cited. The following are some descriptions of the categories of people.

7.1 FLIGHT CONTROLLERS

The term "flight control" initially referred to those parts of an aircraft that control it in flight, such as the rudder, flaps and ailerons. The term "flight test" was at the core of the Langley Memorial Aeronautical Laboratory, where Christopher Kraft spent the first thirteen years of his professional career, in what was known as the "Flight Research Division." The term "flight controller" was used in Mercury as part of Kraft's 1959 concept of how his team could help the astronaut, especially during critical phases of flight and when the "capsule" was over remote sites. By definition, a flight controller is a member of the flight control team that monitors the spacecraft and its orbit and makes the decisions and takes the actions that are necessary to undertake the mission safely and successfully. It is the flight control team's paramount responsibility to protect the crew and bring them home safely.

Even before the first Mercury flights, all the original "flight control" positions were defined and the training of the people who were to man them had begun. In addition to the flight controllers in the Mercury Control Center, for any Mercury mission there was an average of thirteen teams at remote sites. A remote site had at least three flight controllers: a CapCom, a Systems Engineer, and an Aeromed. The Bermuda site could have up to ten flight controllers and an astronaut or two. Thus up to forty flight controllers would be deployed around the world. All of these men went through some form of flight controller training. And of course there were thousands of recovery people.

During a mission, there were arrangements for staff at the Space Task Group and other engineering organizations at Langley to serve as flight controllers on a temporary basis. This worked better in some cases than in others. Eventually the requirements of time and travel for simulations, launch site test support, and real time mission support led to the decision to use dedicated personnel assigned full time to the flight control organization. These first flight controllers were a mix of NASA, contractor, and military medical people.

Today, most MOCR/FCR flight controllers are selected and trained for full time positions but many people in the support areas and the rooms that support the missions are present on a temporary basis, being called in when their particular expertise is needed.

As global communications improved, the number of teams deployed to remote sites was gradually reduced. Even so, seven teams with six flight controllers each were deployed for the Gemini missions. After the unmanned Apollo LM flight in January 1968 these were gradually phased out.

During the Apollo lunar landing missions, the flight controller positions in the MOCR 2 were typically as follows:

- Flight Director.
- Assistant Flight Director.
- Flight Activities Officer.
- Flight Dynamics Officer (FIDO).
- Retrofire Officer (RETRO).
- Guidance Officer (GUIDO) (Kraft preferred the call out GUIDANCE).
- Lunar Module/Telemetry, Electrical, EVA Mobility Unit (LM/TELMU).
- Lunar Module/Control (LM/CONTROL).
- Portable Life Support System-1 (PLSS-1).
- Instrumentation/Communications Officer (INCO).
- Operations & Procedures Officer (O&P).
- Flight Surgeon (SURGEON).
- Command & Service Module/Guidance Navigation Control (GNC).
- CSM Environmental, Electrical, Consumables (EECOM).
- Network Controller (NETWORK).
- Capsule Communicator (CapCom).

Sometimes a console would serve multiple purposes, depending on the phase of a mission. For example, the Booster Systems Engineer would be present for a launch and later the console would be used by the Experiment Activities Officer for the deployment and/or operation of an experiment. Sometimes the Orbital Science Officer would be seated there, representing the Science/ALSEP SSR.

The back row of the MOCR was occupied by people who weren't regarded as flight controllers, but managers of their respective organizations. They certainly had flight responsibilities but not in the sense of "control" related to the conduct of the flight. In Mercury, Walt Williams held the position of Mission Operations Director. He and Flight Director Christopher Kraft would have had a great many discussions about the progress of a mission but the ultimate decision was always the Flight Director's; just as it is now, nearly 60 years later.

The people in the back row included:

- Public Affairs Officer (PAO) who was the mission Commentator.
- Director of Flight Operations (FOD).
- Department of Defense Representative (DOD).
- Headquarters Representatives, such as Administrator, Program Manager, and Mission Director.

Other people were also present in the MOCR at different phases of a mission. There were the various position assistants, shift handover people, TV controllers, photographers, and some technical representatives from the Staff Support Rooms who periodically came in for briefings.

The MOCR flight controllers could call upon a great many support engineers, systems analysts and technicians. These people were located in the Staff Support Rooms adjacent to the MOCR, as well as rooms on the first floor that were in the RTCC and CCATS areas. These people didn't receive flight control training, but they often participated in the training simulations. Similarly, there were analysts and engineers in other buildings, in particular in the Mission Evaluation Room in Building 45 and the Computer Auxiliary Room in the Administrative Wing A of Building 30 who would provide support on demand.

Flight controllers are highly trained and educated. Today, a young person will get more applicable training in college than was available for the early programs because there are more directly applicable courses in aerospace and aeronautical engineering. In the early 1960s, colleges didn't have any courses so NASA STG had their own schools with instructors from those companies that were building the Mercury "capsule" and the Redstone and Atlas launch vehicles. NASA also hired contractors to support this training. A flight controller candidate would at least have an engineering or science degree, and preferably also some practical experience. Initially the people who evaluated the readiness of flight controller teams came from the flight control organization. They were just as well trained, and required to know the systems and procedures well enough to test the whole team. An updated formal training course was held in April 1962. This involved 156 hours of classroom lectures on spacecraft and ground systems. The original NASA flight controllers were the instructors, and they were responsible for the training lesson plans. The primary sources of information for this training were the Flight Controllers' Handbook, Mercury Familiarization Manuals, Network Familiarization books, and manuals from the contractors. Over a year a total of six classes were held for general training. Some students received training that focused on the launch phase, others focused on spacecraft systems so that they could serve at remote sites.

Fast forward through Gemini to Apollo, and the concepts for training flight controllers became better defined and organized. By that time the flight control teams had real experience of systems anomalies and mission accidents and this influenced training for Apollo. There were people whose full time jobs were to train flight controllers, and part of that training involved simulating the various flight modes and their potential failures.

By Apollo, the software analysis tools, the options, and the recommendations in areas such as launch phase design, abort mode definition, rendezvous methods, lunar trajectory simulations, lunar navigation methods, use of propulsion systems and guidance for lunar landing, entry techniques, and some consumables analysis were much better defined.

The final step in flight preparation involved closed-loop simulations with the entire flight control team in the MCC and the flight crews in their spacecraft and mission simulators to test and verify the planning, mission rules, and procedures for each of the major phases of the mission. Specialists had to determine how to display, monitor and control all of the essential real time steps in performing the flight controllers' duties. To learn how to handle any conceived situation would require training sessions designed to uncover any potential weaknesses, whether that be man or machine. Some of the simulated failures were very specific, such as an instrument or subsystem failure. Others were general, such as simulating a control center power failure, a flight controller's console failure, or even a flight controller suffering a heart attack. Some simulated failures were thought to be so unlikely as to never occur…but then they did! The teams spent far more time simulating failures than actual time during a mission. History has proven that simulations are a vital contributor to mission success. Dr. Robert Gilruth once said that what he was worried about was not the unknowns but the "unknown unknowns." Simulations have a way of flushing out a lot of the problems with how a team functions in a complex environment, and they are fundamental to flight controller training.

When outlining "What Made Apollo a Success?" Flight Director Eugene F. Kranz and James Otis Covington provided the following reasons (summarized and paraphrased):

- One of the basic Flight Control Division philosophies is that operations personnel take part in planning a mission from its conception through its execution. Their operational experience contributes greatly in the early stages of mission design and in setting basic design requirements for the various spacecraft systems. As a result, both the spacecraft hardware and mission design have optimum operational qualities.
- Concurrent with the development of operations concepts and mission guidelines and constraints, the flight controllers begin to gather detailed systems information from the spacecraft manufactures and they translate functional schematics and engineering drawings into personal handbooks of spacecraft systems. When used as training tools, these lead to insights into normal and contingency situations, and assist in the development of mission rules and operational procedures.
- Six distinct techniques were used in training Apollo flight controllers, in some cases simultaneously:

  1. Documentation development.
  2. Intimate knowledge of hardware and software.
  3. Formal classes.
  4. Programmed instructional courses.
  5. Cockpit system trainers.
  6. Simulations.

Almost invariably, the simulations changed the mission rules and procedures related to real time decision making, and these resulted in changes to the formal documentation, in particular Crew Checklists, Flight Control Handbooks, Flight Controller Console Procedures, etc. Simulations could even prompt changes to a flight controller's displays, and thereby the software that drove those displays.

For the past 40 years, the operating concept of a ground-based flight control team has been very effective. This envisioned the team responding to program objectives and being staffed with all the disciplines for making decisions in real time. The concept has been driven by the need for rapid and critical response to flight situations, often providing a response within just a few seconds. This has led to careful selection of personnel and a widely scoped training regimen from classrooms to fully integrated simulations with the astronauts in simulators. An early emphasis was established for flight controllers to develop as many of their own tools as possible, from functional schematic systems handbooks to mission rules and procedures, and to the requirements for all of the processing activities, displays, and controls on an individual console. And each console operator was supported by colleagues in staff support rooms and, if necessary, at contractors' facilities. With variations for program content and mission phases, this concept and the makeup of the flight control team has been remarkably consistent.

Nowadays there are formal steps to becoming certified as a flight controller. Training is a phased process, starting in the classroom and proceeding through individual systems training in special facilities and interactive computer-based training. In some cases, it is even contracted out. It is now focused on the ISS, which requires 24/7 flight controller support in many more areas as a result of being an international effort. You can even watch this training on YouTube.

7.2   SIMULATION PERSONNEL

The people that trained flight crews and flight controllers were a special breed. They delighted in honing individuals and teams to a sharp edge by highlighting their weaknesses. But this was welcomed because it emphasized the validity of correct decisions or optional paths designed to circumvent the presented failure mode or situation. As a methodology it was brutal, since it sought to expose an individual's mishandling of a dire situation, and in some cases an entire team's mishandling of the situation. Considering the complexity of a lunar mission and the many different phases of operations, the scope for making a mistake is very great. As a result, many simulations are needed to flush out the most critical of decisions for the most dangerous of operations.

Because the simulations included crew simulators as well as simulations of operations, the simulation team consisted of hardware and software engineers, mission analysts, systems people and technicians. Many simulations were to test complex maneuvers in Earth orbit or lunar orbit. Others mimicked loss of power to the control center, the severing of a critical data cable, or the malfunction of a console. These things have actually happened. The people who thought up these failures to test the responses of the flight controllers and astronauts where indeed "devilishly clever," but their sheer ingenuity was not always appreciated. After a simulation, a flight controller would often complain to the simulation supervisor that the situation with which he had been presented simply wasn't realistic. This was not so often the case between the astronauts and the instructors who ran the crew simulator. But unlikely things can, and do, go awry in flight.

In between missions, the people that devised the simulations wrote software to drive the crew simulators and the flight controllers' displays. The development of the Apollo simulation programs and the associated trainers closely paralleled that of the flight hardware and the increasing complexity of missions. As operations advanced from Gemini to Apollo with more than one spacecraft, and the mission phases moved from Earth orbit to lunar orbital activities and to the lunar landing, the scope and capability of simulations matured to keep pace with the increasing complexity.

Gemini provided an excellent start for Apollo training, because its progress in accurately simulating the launch, rendezvous, and entry modes would carry over. In fact the first Apollo part-task simulators were adapted Gemini simulators. The computer complex and infinity-optics system developed for the Gemini mission simulators was later integrated with simulated crew stations when producing the CM procedures simulator (CMPS) and the LM procedures simulator. (It was the many boxes stuck onto the spacecraft simulator that prompted the nickname of a "train wreck.") For the translation and docking simulator, the simulated Agena target vehicle and Gemini spacecraft were replaced with the CM target mockup and the LM crew station.

No fewer than fifteen simulators trained crews during the Apollo years. Three were the primary Command Module Simulators; one at Houston and a pair at the Cape. Two were the primary Lunar Module Simulators, one at each location. At Houston a Command Module Procedures Simulator was used just to train crews to rendezvous with the CM, as there was a Lunar Module Procedures Simulator for LM rendezvous and landing training. The Gemini Dynamic Crew Procedures Simulator was adapted to train Apollo crews for launch and launch aborts. Other moving-base simulators at Houston were for lunar module formation flying and docking. And there was also a centrifuge (to avoid visits to Johnsville). Langley pioneered the research into the final 200 feet of lunar descent by building a rig that balanced five-sixths of a simulator's weight in order to allow astronauts to practice controlling the LM in the gravity of the Moon. Another lunar landing simulator used a jet engine to support five-sixths of its weight and permit free-flight training of the final phase of the descent. In fact, that simulator needed a simulator of its own to prevent the astronauts from crashing it. There were two partial-gravity simulators that used rigs to allow the astronauts to walk in space suits with five-sixths of their weight cancelled. And the Marshall Space Flight Center created an Earth-gravity version of the Lunar Roving Vehicle to enable astronauts to rehearse the traverses which they would undertake on the Moon.

Figure 7.1 A Command Module Simulator. Photo courtesy of NASA.

Figure 7.2 The Instructor's Console of a CM Simulator. Photo courtesy of NASA.

Figure 7.3 A Lunar Module Simulator. Photo courtesy of NASA.

Figure 7.4 Neil Armstrong in the Lunar Module Simulator at KSC on June 19,1969, less than a month before the launch of Apollo 11. Photo courtesy of NASA.

Among the plethora of simulators, use of the Command Module Simulators and Lunar Module Simulators nonetheless consumed 80 per cent of the Apollo training time of 29,967 hours. These simulators, and their associated computer systems, were crucial to the success of the program. The emergency that faced Apollo 13 in 1970, proved the high fidelity and flexibility of the simulators by enabling all lunar module engine burns, separation events and maneuvers to be evaluated in advance and ad hoc procedures developed as the crippled mission progressed.

The procedures simulators were driven by a single mainframe computer, but the Mission Simulators employed networks of machines. Honeywell received a $4.2 million contract on July 21, 1966, to supply DDP-224 computers for these complexes. Singer-Link was once again the contractor for the simulators. They used three computers for the Command Module Mission Simulator and two for the Lunar Module Simulator. The sets of computers could communicate among themselves by using 8K words of common memory, in which the data required across the simulation would be stored. Later, a third and fourth computer were added to the LM and CM simulators respectively. These simulated the onboard computers. By the time of the Apollo 10 training, a fifth computer filled out the CM Simulator by mimicking the Saturn vehicle. When the MCC was hooked up to the spacecraft simulators it was possible to run integrated simulations with up to ten digital computers working on one large problem simultaneously.

Software became as essential to the simulated world of Apollo as it was in the real world. Software development for the Apollo mission simulators required the efforts of 175 programmers at the peak, compared to 200 hardware persons. Over 350,000 words of programs and data eventually ran on the two simulators. Simply by recording the status of the computers and data on magnetic tape and reloading memory to match the state of the software at a given time in the mission, trainers could return the crews to a certain point in a simulation. This flexibility made the training task much easier.

So there were hundreds of simulation people involved in Apollo, owing to the number of simulators spread across NASA and the complexity of the operations. These personnel were of every type, including hardware and software engineers, computer programmers, technicians, and instructors.

In contrast to the relatively private training by astronauts in crew procedures trainers, simulations in which the astronauts were linked in to the MOCR for an integrated simulation were "open," possibly even with members of the public in the viewing area of the MOCR. The debriefings after a simulation was complete were heard by everyone on the communications loop, whether in the MOCR, an SSR, or almost anywhere in the building (or around the world in other facilities providing support). Mistakes were inevitable. But they were not treated as cause for punishment, they were lessons learned. We were being judged by the Flight Director and our peers. This was the simulation world for spaceflight operations people, and it produced teams sufficiently skilled to enable missions to fly to the Moon and return.

Some of the "unsung heroes" of the simulation world, in particular those who came up with the original concepts and methodologies of mission simulations for Project Mercury and went on to implement them for Gemini and Apollo, include:

- Harold Miller.
- Richard Hover.

- Dick Koos.
- Glynn Lunney.
- Jack Cohen.
- Gordon Ferguson.
- Jim Miller.
- Carl Shelley.
- Gerald Griffith.
- Peter Segota.
- Efren Cavillo.
- Miles Springfield.
- Mel Brooks.
- Lee Roots.
- Richard Brown.
- Ray Zedeker (crew training).

In the words of Chief, Flight Crew Operations Division Warren J. North, and Chief, Crew Training and Simulation Division Carroll H. "Pete" Woodling: "The Apollo flights speak for the value of this simulation effort in verifying late changes, validating procedures, and establishing crew readiness."

7.3 COMPUTER SUPPORT

All of the computers in Building 30 were mainframes, occupying entire rooms. Personal computers as we know them today were barely imaginable. The IBM and UNIVAC contractors supported the mainframes on the first floor where the RTCC and CCATS were located, but NASA and their support contractors wrote the operational software that was sent to the contractors for implementation.

And systems such as the IBM 7094 in the Auxiliary Computer Room on the third floor of the Administrative Wing of Building 30 and the UNIVACs in the MSC Central Data Facility in Building 12 which supported the MCC were also mainframes.

Discipline specialists inside MPAD developed trajectory planning techniques and the tools utilized in mission planning and analysis, both for nominal and for contingency/alternate missions. Software was created for rendezvous planning, lunar trajectory design, spacecraft consumables management, ascent trajectory planning, abort and contingency planning, orbit determination, etc. Most of the functionality required for pre-flight planning and analysis was later used by the RTCC for real time operations and for simulation support. When planning tools were verified, the trajectory specialists sent the software logic, formulation, and code to IBM for implementation in the RTCC.

The following are just a few examples:

- The Rendezvous Analysis Branch developed the tools for Earth orbit and lunar orbit rendezvous planning. As the tools were verified, Bill Sullivan was responsible for documenting the package that was passed to IBM for implementation.
- Emil Schiesser and his group developed the tools for orbit determination, trajectory propagation, ground navigation, and geopotential models of the Earth and Moon in the RTCC.

- The Math Physics Branch generated continuous improvements from one mission to the next in lunar navigation. As tracking data was accumulated and analyzed, there was increased knowledge of the mapping of the lunar mass concentrations ("mascons") that perturbed the orbital trajectories of spacecraft. This enabled navigation accuracy to be improved.
- The Math Physics Branch provided standardized coordinate systems, and transformations from one coordinate system to another coordinate system for implementation in the RTCC and other organizations across NASA.
- Jerry Yencharis and his team developed the real time applications for the translunar midcourse corrections, lunar orbit insertion, returning to Earth, etc. These were provided directly for implementation in the RTCC/ACR. He also formulated the S-IVB translunar injection guidance equations and the MSFC Hypersurface.
- MPAD provided simulation data packs to support all mission simulations (for both nominal and contingency planning) and provided support in the ACR and SSR.

See Section 7.6.1 Mission Analysis for how the various Mission Planning and Analysis Division Branches developed the initial software for the RTCC that was sent to IBM for implementation.

7.4   RECOVERY PERSONNEL

Given the sheer size and scope of the NASA spacecraft recovery program and the number of DOD personnel involved, it would be difficult to single out key people, especially those that could be considered as "pioneers." It is clear that most such people came from the Langley Research Center and the Space Task Group. Even in the early 1959 concept phases of Project Mercury the challenge of recovering a spacecraft that could land anywhere in the world was a daunting one. Dr. Robert Gilruth assigned the task to Chief, Operations Division Charles W. Mathews. He enlisted the help of Robert F. Thompson, who had some Navy experience.

Thompson's Branch was involved in qualifying the capsule for landing on land as well as at sea. Drop tests were conducted to evaluate the landing bag as well as to verify the capsule would float properly. The various recovery aids were tested, such as dye markers, shark repellents, flashing lights, and radio locators. Branch personnel were involved with the design of recovery systems, manning of Navy recovery vessels, Air Force aircraft, and coordinating with the mission planners about the planned recovery positions. Once the spacecraft was in orbit, Thompson coordinated with the Navy regarding possible and planned capsule landing points.

This Branch included the following people from Langley/STG (in alphabetical order):

- Peter Armitage (former AVRO engineer).
- Gilbert M. Freedman.
- John B. Graham.
- Harold E. Granger.
- Jerry Hammack.
- William C. Hayes.

- Leon B. Hodge.
- Enoch M. Jones.
- John C. Stonesifer (came onboard in 1961).
- Robert F. Thompson.
- Charles I. Tynan, Jr.
- Milton L. Windler.

After Mercury and the relocation to Houston, the MSC organization changed to accommodate Gemini and Apollo but the job of recovery, no matter the size, was still a daunting one that had to evolve.

In the new 1962 MSC Gemini organization, the new Landing and Recovery Division was headed by Robert F. Thompson. The Division remained a part of Flight Operations, although in view of the nature of the task one could consider adding Fleet Operations. Peter Armitage was Chief, Operation Evaluation and Test Branch in the Division. He had previous Search and Rescue experience as well as ejection seat and parachute testing, and also experience with developing and qualifying the electronic location aides and astronaut survival training. He continued his recovery engineering responsibilities through Gemini and into Apollo. William Hayes was Chief of the Operations Branch which determined ship and aircraft positions and procedures for world- wide recovery operations.

On moving to Houston, Jerry Hammack became Deputy Manager, Vehicles and Missions. He was made Chief, Future Programs Division in 1966, then in 1971 he became Chief, Landing and Recovery Division. From 1973 through to 1987 he was Chief, Safety Division. He therefore led the recovery activities for many missions.

John Stonesifer started off at Langley but in view of his Navy experience he was hired by the STG Recovery Branch. He assisted in planning and preparing documentation requirements for positioning ships and aircraft for John Glenn's MA-6 mission. During that, Stonesifer assisted Thompson by working with the DOD officer commanding the recovery units from the Mercury Control Center. For the Mercury, Gemini, and early Apollo missions, Stonesifer assisted and/or directed all the phases of spacecraft operational planning, the development and testing of the unique hardware required for recovery, and the training of DOD world-wide ship and aircraft forces. During this time, he was the NASA Team Leader and technical advisor on a number of the primary recovery ships.

Dr. Donald E. Stullken joined NASA from the Naval School of Aviation Medicine (NSAM) in Pensacola, Florida. He was Chief of Recovery Operations Branch for the later Mercury missions, and then for Gemini, Apollo, Skylab and Apollo-Soyuz. As the senior NASA Team Leader aboard the primary recovery ship, he was often the first person to welcome the crew when they stepped from the recovery helicopter. While at the NSAM, he initiated the development of the "Stullken floatation collar" that helped to stabilize and buoy a capsule after it had splashed down. In the late 1950s, he was also a member of the team of scientists that developed the biocapsules in which the monkeys were sent on high ballistic trajectories by U.S. Army Jupiter rockets. He helped to create the life support system, animal restraints, and animal training.

During the later expansion of the Recovery organization, and for subsequent programs, Harold Granger led the advanced planning for the worldwide recovery support when it became necessary to change the positioning of ships and aircraft in response to world events. Timely coordination with the DOD to determine the availability of its forces were important.

Milton Windler's early work involved developing and qualifying advanced electronics for location aides in recovery operations, and he also worked on the training of astronauts and recovery personnel.

7.5 SCIENCE SUPPORT

There is a direct connection between what scientists in 1962 initially expected the astronauts to do on the Moon and the type and number of scientists that helped to design the scientific experiments, and those that manned the Science/ALSEP SSR years later during the lunar landings. Although Apollo was a political program to make America the first nation to reach the Moon, many people wanted the crews to carry out science on and around our nearest celestial neighbor. However, some NASA managers and engineers regarded science as a distraction from the already daunting engineering challenge of landing a man on the Moon and returning him safely to Earth.

Once President Kennedy announced the goal, NASA Headquarters appointed physicist Charles P. Sonett of Ames Research Center to lead an ad hoc working group to outline an Apollo science program. The Office of Manned Space Flight at Headquarters provided some guidance on the present thinking concerning the program. The group's twelve members and nine consultants included geologist Eugene Shoemaker, geophysicist Paul Lowman, astronomers Thomas Gold and Gerard Kuiper, and chemist Harold Urey provided their draft report to NASA in July 1962 and circulated it at the National Academy of Science's ten-week Iowa City meeting that summer. Bear in mind that this was as Project Mercury started orbital flights and construction of the Houston MSC was underway, so not many people were thinking about Apollo.

Briefly, the group proposed the following:

- An automated Lunar Orbiter should precede the flights to make the final landing site selections.
- The Apollo crews should include scientist-astronauts with doctorates and 5–10 years' experience in the field.
- Graduate students should be trained as potential astronauts for the future.
- Pressure suits should be designed to be as flexible as possible in order to facilitate lunar field work.
- There should be the capability to reach areas as far as 50 miles from the landing site. (Headquarters had suggested the possibilities of a rover.)
- There should be an automated supply lander with up to 15 tons of cargo for a modest scientific program.
- A list of experiments should be compiled that included the collection of lunar material. (They discussed the contingency sample.)
- There should be two lists of different candidate landing sites, with a total of 28 sites. (These were based upon terrestrial observations.)
- Landings should be made in equatorial, polar, and other selected areas.

The report went into more detail than summarized here. All these objectives were refined over the next couple of years with the participation of hundreds of scientists. A look at the Apollo Experiment Program Plan lists literally hundreds of scientists from universities across the U.S., one from Switzerland and another from Australia. Also listed were the NASA scientists and contractors that would build and test the experiments. There would be a Principal Investigator for each experiment. He would be supported by at least one Experiment Manager and a Project Engineer. There might be Co-Investigators for different missions. Each team would have supporting scientists and graduate students.

It would not be practical in this book to list all the scientists that supported Apollo. The people that supported the Science/ALSEP SSR are in the Mission Manning Lists of Appendix 2.

The following were the Principal Investigators for the ALSEP experiments, some of whom staffed the SSR:

Lunar Surface Experiments Investigators

- Passive Seismic Experiment (S-031): Dr. Gary Latham, Columbia University.
- Active Seismic Experiment (S-033): Dr. Richard L. Kovach, Stanford University.
- Stanford Lunar Seismic Profiling (S-203): Dr. Richard L. Kovach, Stanford University.
- Lunar Portable Magnetometer (S-198): Dr. Palmer Dyal, NASA Ames.
- Magnetometer Experiment (S-034): Dr. Charles P. Sonett, NASA Ames.
- Solar Wind Experiment (S-035): Dr. C.W. Snyder, JPL.
- Suprathermal Ion Detector (S-036): Dr. J.W. Freeman, Rice University
- Heat Flow Experiment (S-037): Dr. Marcus Langseth, Lamont Geological Observatory.
- Charged Particle Lunar Environment Experiment (S-038): Dr. B.J. O'Brien, Rice University.
- Cold Cathode Gauge Experiment (S-058): Dr. F.S. Johnson, University of Texas.
- Field Geology Experiment (S-059): Dr. E.M. Shoemaker, CalTech, Dr. G.A. Swann, USGS and Dr. W.R. Muelhlberger, University of Texas.
- Laser Ranging Retro-Reflector (S-078): Dr. James E. Faller, Wesleyan University and Dr. Carrol Alley, University of Maryland.
- Solar Wind Composition Experiment (S-080): Dr. Johannes Geiss, University of Bern, Switzerland.
- Cosmic Ray Detector (S-151): Dr. Robert Fleischer, General Physics Laboratory.
- Lunar Portable Magnetometer (S-198): Dr. Palmer Dyal, NASA Ames.
- Far Ultraviolet Camera/Spectroscope (S-201): Dr. T.L. Page, NASA MSC and Dr. Carruthers, Naval Research Laboratory.
- Dust Detector (M-515): Dr. Brian J. O'Brien, University of Sydney, Australia.
- Cosmic Ray Detector (S-152): Dr. R.L. Fleischer, General Electric, Dr. R.M. Walker, University of Washington and Dr. P.B. Price, University of California.
- Lunar Ejecta and Meteorites (S-202): Dr. Otto Berg, NASA Goddard.
- Soil Mechanics Investigations (S-200): Dr. James Mitchell, University of California.
- Lunar Gravity Traverse (L-069): Dr. C.G. Wing, MIT.
- Surface Electrical Properties (S-204): Dr. Gene Simmons, MIT, Dr. David Strangway, University of Toronto, Canada.
- Lunar Atmospheric Composition (S-205): Dr. John Hoffman, University of Texas at Dallas.
- Lunar Surface Gravimeter (S-207): Dr. Joseph Weber, University of Maryland.

- Lunar Neutron Probe (S-229): Dr. D.S. Burnet, CalTech.
- Dust Thermal Radiation Engineering Measurement (M-5-5): James B. Bates, NASA JSC.

Lunar Orbital Experiments Investigators

- Gamma Ray Spectrometer (S-160): Dr. J. R. Arnold, University of California.
- X-Ray Fluorescence (S-161): Dr. I. Adler, NASA Goddard.
- Alpha Particle Spectrometer (S-162): Dr. Paul Orenstein, Smithsonian Astrophysical Observatory.
- S-Band Transponder (S-164): W.L. Jorgen, JPL.
- Mass Spectrometer (S-165): Dr. John Hoffman, University of Texas at Dallas.
- Far UV Spectrometer (S-169): Dr. W.G. Fastie, Johns Hopkins University.
- Bistatic Radar (S-170): H.T. Howard, Stanford University.
- IR Scanning Radiometer (S-171): Dr. F.J. Low, University of Arizona and W.W. Mendell, NASA JSC.
- Particles and Shadows Subsatellite (S-173): Dr. K.A. Anderson, University of California.
- Particles and Fields Magnetometer Subsatellite (S-174): Dr. R.J. Coleman, Jr. UCLA.
- Laser Altimeter (S-175): Dr. William M. Kaula, UCLA, W. Sjogren, JPL.
- Lunar Sounding Radar (S-209): Dr. Roger J. Phillips, JPL, Dr. Stanley Ward, University of Utah, Walter E. Brown, Jr., JPL.
- Lunar Multispectral Photography (S-158): Dr. A.F.H. Goetz, Bellcomm.
- Electromagnetic Sounder "A" (S-168): Dr. S.H. Ward, University of California.
- Apollo Window Meteoroid (S-176): R.E. Flaherty, NASA MSC.
- UV Photography, Earth/Moon (S-177): T. Owen, IIT Research Institute.
- Gegenschein Photography (S-178): L. Dunkelman, NASA Goddard.
- Solar Wind Spectrometer: D. Lind, NASA MSC.

These scientists were supported by the Science & Applications Directorate headed by Robert Piland, the Lunar Science Institute headed by Dr. William W. Rubey and the Apollo Program Office headed by George Low. There were also representatives in the SSR as well as in the SPAN and MER. Operationally, the Flight Activities Officer in the MOCR would provide the interface between the scientists and the crew. The INCO officer would facilitate the communications with the crew and the experiments.

Scientists and engineers from the manufacturers of the experiments were also supporting their Principal Investigators. These companies included the ALSEP prime contractor Bendix Systems Division and subcontractors, Martin, General Electric, Philco, Teledyne, and other equipment providers.

7.6   SYSTEMS ANALYSIS SUPPORT

7.6.1   Mission Analysis

The analysts that that supported the Apollo Mission Control Center came from several different NASA MSC organizations as well as Apollo contractors. They supported pre-mission planning, on-going operations, and post-flight analysis. In the mission planning phase, the various MSC organizations and their contractors each had specific systems analysis staffs. Many of the trajectory systems analysts that supported the Apollo missions came from the Mission Planning and Analysis Division (MPAD) that was formed in 1959. It was initially headed by John Mayer with Howard Tindall as his Deputy through to May 1969. After Mayer, Ron Berry was the Division Chief until March of 1985. Ed Lineberry and Claude Graves led the group until it disbanded in 1990. By then it had supported everything from Mercury, Gemini, Apollo, ASTP, Skylab and Shuttle through the International Space Station.

A good example of Apollo pre-mission analysis spanning flight dynamics, trajectories and rendezvous techniques were Tindall's efforts as Chief, Apollo Data Priority Coordination. His responsibilities were wide ranging. He chaired meetings between astronauts, flight controllers, MPAD, design engineers, MIT, TRW, and other contractors. Such meetings were often contentious but Tindall had a way of adjudicating disagreements and overseeing the details of planning mission techniques. His efforts impacted the software in the ACR, where some methods were tested, and eventually in the RTCC for the operational software. Go to www.tindallgrams.net for a list of his insightful memos.

Many trajectory analysts "cut their teeth" on the development of the tools to compute trajectories of all types, and for all phases of flight. These tools were basically prototypes for the software subsequently implemented by IBM in the RTCC. As a result they became analysts of off-nominal situations and during a mission they would man the Flight Dynamics SSR and ACR to offer support to the MOCR flight controllers in the event of a problem arising.[1]

In addition to John Mayer and Howard Tindall, the following are considered the pioneers of mission planning and analysis:

- Carl Huss: Developed trajectory analysis for launch, orbit and entry, and related Mission Rules. Devised nominal and contingency trajectories and abort criteria. The first Retrofire Officer (RETRO).
- Ted Skopinski: Developed early expertise in orbital mechanics. Co-author with Carl Huss and John Mayer of the first textbook on Space Technology written in early 1958, well before the formation of the STG.
- Emil Schiesser: Development of an Orbit Determination capability for the RTCC and ground navigation capability. Made major contributions to the refinement of the terrestrial and lunar geopotential models that were used by planning software in the RTCC.
- Mary Shep Burton: Originator and manager of Math Aides in the MPAD. She developed the early concepts of data reduction and planning product documentation and configuration management. The Math Aides function was absolutely essential to the success of the MPAD.
- Ed Lineberry: Instrumental in the development of rendezvous techniques and rendezvous software. Specifically, Lineberry developed the logic and equations for the

---

[1] A good example of the lunar trajectory software required for Apollo Lunar Rendezvous can be seen by going to: https://www.clipzui.com/video/r3p3w5f3h365y2p4e4w513.html. Although an old video, it visually explains what calculations are required and why. Another good video of the Apollo Guidance Computer is: https://www.clipzui.com/video/r3m3t494w235k4g49544x2.html.

advanced Analytic Ephemeris Generator essential to the development of rendez-vous planning tools and the rendezvous capability in the RTCC. He is also known for his formulation and documentation of the Jacchia-Lineberry Upper Atmosphere Density Model.

- Lyn Dunseith: He began his career at Langley. Later he was instrumental in the functional design and development of the system architecture in the RTCC. As Chief, Real Time Program Development Branch he developed the requirements for the real time software that IBM would implement in the Mission Control Center.
- Clay Hicks: He came from the Langley PARD. With John Mayer, Clay was one of the first four members of the Mission Analysis Branch of the STG. He was instrumental in the conceptual design of the first Mercury trajectories. He was a specialist in launch trajectories, launch aborts, g-loads, and reentry heating.
- Hal Beck: Instrumental in the development of the lunar trajectory design capability, including the early iterative methods for targeting translunar trajectories. He also managed the development of the Apollo Reference Mission Program, essential for the development of the Apollo Reference Mission for each lunar mission.

Figure 7.5 MPAD moved to MSC in 1964. L-R: the women are Doris Folkes, Cathy Osgood, Shirley Hunt, and Mary Shep Burton; the men are Dick Koos, Paul Brumberg, John O'Loughlin, Emil Schiesser, Jim Dalby, Morris Jenkins, Carl Huss, John Mayer, Bill Tindall, Hal Beck, Charlie Allen, Ted Skopinski, Jack Hartung, Glynn Lunney, John Shoosmith, Bill Reini, Lyn Dunseith, Jerry Engel, Harold Miller, and Clay Hicks. Seventeen members of the team are not present: Ed Lineberry, Marty Jenness, Joe Thibodaux, Jim Ferrando, Marlowe Cassetti, Jerry Bostick, Don Incerto, Gerry Hunt, John Gurley, Jim Rutland, Myrtle Braslow, John Lewis, Athena Markos, Fran Johnson, Pattie Leatherman, Nancy Carter, and Edna Hawkins. Photo courtesy of NASA and Hal Beck.

A good example of how MPAD worked with the MOCR flight controllers and the IBM and Philco WDL contractors is given by Neil Hutchinson, who was then in the Systems Logic and Processing Section and subsequently became an Apollo Flight Director. He interfaced with the systems flight controllers to find out what real time data from launch vehicles and spacecraft they felt was essential to view on their consoles. This information was turned into requirements documents and given to IBM for coding. Once completed, the software was evaluated by full up testing, with the MOCR flight controllers evaluating their displays (which were cathode ray tubes, event modules, and clocks). Revisions were made as required until the flight controllers were satisfied.

See Appendix 2 for the Mission Manning Lists which lists other people who supported the missions.

7.6.2 Spacecraft Systems Analysis

Spacecraft systems analysts came from various Divisions and Branches of the Engineering and Development Directorate, the Apollo Systems Program Office, and the contractors North American Rockwell, Grumman Aircraft Corporation, Bendix, MIT and others. They covered every system and subsystem of the CSM, the LM and even the crew's EVA systems and suits. They also included systems related to the Lunar Roving Vehicle. And there were people from Reliability and Quality Assurance. The Science and Applications Directorate provided analysts for the ALSEP packages and other experiments.

While a systems analyst may well have been an engineer during the design or manufacturing phases, those who supported flight operations were focused more on how things worked, and they were ready to analyze any that failed or did not function as planned. The systems analysts would assist the flight controllers with off-nominal situations, failure analysis, and work arounds to the potential failures. They might normally be assigned to an engineering organization but attend flight operations meetings prior to a mission, during flight, and during the post-mission analysis. Their mission operations support was addressed under sections 5.3 Staff Support Rooms and 5.4 Other Support Rooms which described the SPAN and the MER.

The Mission Manning Lists in Appendix 2 name many people that supported the SPAN/MER activities, but the following people, some of whom were flight controllers, stand out as pioneers in the field of systems analysis and testing. In truth, whole chapters could be written about these individuals but here are some brief summaries:

- Walter Kapryan: An original STG member, he was one of the first Cape MCC Mercury systems experts. Later he became the Director of Launch Operations for Apollo and the Shuttle.
- Don Arabian: An expert of spacecraft systems. In leading the MER and SPAN teams he conceived and developed the Skylab parasol that saved the Orbital Workshop. Later on, he was Manager of Flight Safety for the Shuttle.
- Scott Simpkinson: An original STG member, he managed the Mercury spacecraft systems checkout at the Cape. He managed the assembly and test of the Mercury Atlas booster and then test operations for Gemini. In later years he managed Apollo

flight safety and mission evaluations. He was Manager of Flight Safety for the Shuttle.

- Arnie Aldrich: He was an original STG Mercury systems expert, remote site and MCC systems analyst, and a flight controller. After chairing the Shuttle Orbiter Avionics Software Control Board he became the NASA Associate Administrator for Space Systems Development. Arnie has the unique distinction of being the only flight controller who sent a command to reenter a spacecraft. This was when he saved Enos on the MA-5 flight.
- Aaron Cohen: Manager for the Apollo Command & Service Modules. He oversaw the design, development, production and test flights of the Space Shuttles as Manager of the Space Shuttle Orbiter Project Office from 1972 to 1982. He later became the Director of Engineering and Director of JSC.

Of course, these pioneers had a lot of help along the way. Others with unique systems analysis skills were:

- Sy Rubenstein of North American Rockwell.
- Tom Kelly of Grumman Aircraft.
- Kenneth Cox of the JSC Engineering and Development Directorate.
- Phillip Shaffer, a former FIDO and Flight Director, who worked on the Apollo, Skylab, and Shuttle programs.
- Richard Parten, Chief of the Spacecraft Software Division.
- Robert Ernull, an original member of the STG who worked on the MCC operations and went on to become Deputy Director for Mission Support and worked on Shuttle Flight Software Design Systems and the Center Computer Complex.
- Henry Pohl, Shuttle propulsion systems engineer.
- Joe Thibodaux, the original STG and MSC Chief of Propulsion and Power Division.
- John Hanaway, who, as Chief, Avionics Systems Engineering Division, developed the Orbiter avionics hardware and also the Shuttle Avionics Integration Laboratory (SAIL).

The above are representative of the people at NASA who can truly be called (among other things) spacecraft systems analysts.

7.7 MANAGEMENT

In keeping with the context of the Apollo Mission Control Center, the following people oversaw the initial operational design of the MCC-H facility. While many others tended to the procurement and civil engineering of the necessary facilities in Building 30, the people who drove the requirements from the beginning in the 1962–1964 timeframes were those that would have to operate the control center.

The two pillars of flight operations management were Robert R. Gilruth and Christopher Columbus Kraft, Jr. My favorite picture of them is Figure 7.6, because it says so much. They are not just smiling at each other to celebrate the successful end of the Gemini 5 mission, less than a year after the MCC became operational, they are looking at each other and remembering more than twenty years of work at Langley, the Space Task Group and subsequently the Manned Spacecraft Center which led to this time of celebration. Eleven years older than Kraft, Gilruth was his mentor from at least as far back as the flight test years of the post-war 1940s when Kraft was in his twenties.

Figure 7.6 Chris Kraft and Bob Gilruth celebrate Gemini 5. Photo Courtesy of NASA.

The operational requirements for the control center codified the lessons that were learned in the Mercury Control Center. This specification was supplied to the Western Development Laboratories of the Philco Corporation to design the hardware and associated systems.

In 1962, the entire control center operation was integrated with the Manned Space Flight Network and was referred to in the Philco contract as the Ground Operational Support Systems (GOSS).

The Philco managers were O.B. Schwede, R.S. Cronhardt, and D.A. Linden. The Mercury flight controllers provided inputs via the document called "Flight Control Requirements for the Integrated Mission Control Center." These inputs were coordinated by each Branch in the Flight Control Division and sent to the Flight Support Division, which was canvassing all areas such as the RTCC and CCATS on the first floor. The requirements were then sent to Gilruth and Kraft for approval, then to the Facilities Division, and finally to the Procurement and Contracts Division that prepared the implementation contracts.

Key people responsible for this process included:

- Barry Graves.
- John Hodge.
- Tecwyn Roberts.

- Henry C. Clements.
- Howard Tindall.
- John P. Mayer.
- Eugene F. Kranz.

During the formative period 1962–1964, Walter Williams was the Assistant Director for Operations with three operations divisions under him. Christopher Kraft was Chief of the Flight Operations Division, Merritt Preston was Chief of the Preflight Operations Division, Warren North was Chief of the Flight Crew Operations Division. These latter two Divisions had little impact on the control center design.

While NASA Headquarters did not get involved in the details of the control center design, they certainly saw to its funding and representatives manned the control center during the missions on the upper row of consoles.

Figure 7.7 Key managers Chris Kraft, Samuel Phillips, Chuck Mathews and George Low celebrate the Apollo 11 lunar landing. Photo courtesy of NASA.

The NASA Headquarters managers were responsible for the entire manned spaceflight program. One such was Dr. George E. Mueller, who was Associate Administrator for the Office of Manned Space Flight (OMSF) during the early Apollo era through to Apollo 12. He was instrumental in introducing the all-up testing of the Saturn V that was initially resisted by the team at Marshall Space Flight Center headed by Dr. Wernher von Braun. But in retrospect it is evident that Mueller's test philosophy ensured that we attained Kennedy's challenge of reaching the Moon before the decade was out.

In addition to the all-up testing, Mueller reorganized NASA and brought in a number of people from the Air Force who had managed the development of the latest missiles, most notably General Samuel Phillips. Figure 7.8 shows several NASA managers who ensured the success of Apollo. All of them occupied the back row of the MOCR at one time or another.

Figure 7.8 Apollo 11 managers in the Launch Control Center at KSC celebrate a successful start to Apollo 11 on July 16, 1969. L-R: Charles W. Mathews, Deputy Associate Administrator for Manned Space Flight; Dr. Wernher von Braun, Director of the Marshall Space Flight Center; George Mueller, Associate Administrator for the Office of Manned Space Flight; and Lt. Gen. Samuel C. Phillips, Director of the Apollo Program.

The following is the list of those with mission management responsibility for Apollo 11:

* Lt. Gen. Samuel C. Phillips: Director, Apollo Program, OMSF.
* Maj. Gen. John D. Stevenson: Director Mission Operations, OMSF.
* Lee. B. James: Vehicle Program Manager, Saturn V, MSFC.
* R. Adm. Roderick O. Middleton: Apollo Program Manager, KSC.
* George H. Hage: Mission Director, OMSF.
* Capt. Chester M. Lee: Assistant Mission Director, OMSF.
* Thomas H. McMullen: Assistant Mission Director, OMSF.
* Rocco Petrone: Director of Launch Operations, KSC.
* Christopher C. Kraft: Director of Flight Operations, MSC.

We must acknowledge the first NASA Administrator T. Keith Glennan, and his deputy Dr. Hugh L. Dryden for getting NASA up and running. However we must also credit President John F. Kennedy for listening to Dryden and recently-appointed Administrator James Webb in April 1961 when they provided advice on how to respond to the many Soviet space successes. Some historians credit Dryden with having been the one who spoke up and suggested to the President that a manned flight to the Moon was the way to "catch up" with the Soviets in the space race. Within a month of meeting with Webb and Dryden, on May 25 Kennedy made his speech to Congress where he challenged the nation to land a man on the Moon before the decade was out. From that date on through October 1968, Webb, who knew his way around Washington politics, ensured the space agency received the funding needed to achieve its goals.

NASA top management employed the "program management" concept that centralized authority and emphasized systems engineering. This was critical to enabling management at all levels to achieve the enormously difficult systems engineering, technological, and organizational integration requirements of the Apollo program.

The NASA Administrators during this time were:

- James E. Webb: February 14, 1961 to October 7, 1968.
- Thomas O. Paine: October 8, 1968 to September 15, 1970.
- George M. Low: September 16, 1970 to April 26, 1971.
- James C. Fletcher: April 27, 1971 to May 1, 1977.

7.8   FLIGHT SURGEONS

Since the early days of Mercury, there have been Flight Surgeons in the MOCR and Aeromeds in the SSR. They are generally either MDs or physiologists. The same was the case for the remote sites, when we had them. Flight Surgeons are physicians that have specialized training and board certification in Aerospace Medicine. Many of the original Mercury flight surgeons supported the Apollo missions in the control center. There were others in the Life Systems Branch of the STG Flight Systems Division, which was headed by Dr. Lt. Col. Stanley C. White, that supported astronaut training and provided general medical care such as the astronauts' physician Dr. Lt. Col. William K. Douglas, nurse Dee O'Hara and aeromedical consultant Dr. Robert B. Voas.

On the engineering side, Richard H. Johnston, Dr. Richard Pollard, Dr. Lt. Col. James P. Henry, and Dr. Lt. Col. William S. Augerson were involved with pressure suits and life support systems. Dr. David P. Morris supported crews at the Cape. Others in the initial STG medical organization included Dr. Richard Pollard and Dr. Howard A. Minners.

When it came to flight operations, only the following were listed as Flight Surgeons on the Manning Lists for the Cape Mercury Control Center:

- Dr. Charles A. Berry.
- Dr. A. Duane Catterson.
- Dr. Stanley White.

The monitoring techniques used to observe the critical medical parameters of astronauts in flight were developed by the engineering and operations teams. The flight surgeons' judgment and ability to assess the astronauts' well-being during flight, as well as their confidence in the crew's readiness to undertake a mission, were very necessary to achieving mission success.

During Gemini, the following physicians joined the team which manned the Surgeon console in the MCC:

- Dr. Duane Coons.
- Dr. John J. Droescher.
- Dr. W. Royce Hawkins.
- Dr. G. Fred Kelly.
- Dr. John F. Ziegelschmid.

At the end of Gemini, with 2000 man-hours logged in space, it was clear that man could engage in relatively long space missions without any serious threat to health. But clarification was still required in many areas. First of all, because of the small number of individuals who flew in space and because of the variability of their responses, it was impossible to distinguish between actual space-related physiological changes and individual physiological variations. Secondly, for the changes that were directly related to space, the relatively short mission durations precluded the identification of trends. So four medical objectives were specified for Apollo:

- Ensuring crew safety from a medical standpoint. This required that every effort be made to identify, eliminate, or minimize anything which might pose a potential health hazard to the crew.
- Improving the probability of mission success by ensuring that sufficient medical information was available for management decisions.
- Preventing back-contamination of Earth by lunar material.
- Achieving a deeper understanding of the biomedical changes that result from spaceflight. This objective was formulated to detect, document and understand changes occurring during flight.

Special measures were taken to protect the health and enhance the safety of Apollo astronauts. In addition to a health stabilization program and a pre-flight medical examination, these measures included drug sensitivity testing of each medication that was to be carried aboard the spacecraft.

The pre-flight medical program was designed to preclude, as far as possible, the development of any clinical medical problems during spaceflight. Since no preventive medicine program, however carefully conceived, can guarantee the absence of illness or disease the spacecraft would be provided with a "medical kit," the contents of which would be revised as appropriate during the program. Onboard bioinstrumentation would provide telemetry to monitor vital signs in order to allow rapid diagnosis of any physiological difficulty in a crewmember and provide the medical information needed by mission management functions. Additional information was transmitted via voice communication between the crew and the flight surgeons at MCC. During extravehicular activity, methods were devised to enable metabolic rates to be assessed. In addition to heart rate, oxygen consumption was monitored along with the inlet/outlet temperatures of the liquid cooled garment worn by the crewmen.

Opportunities for in-flight medical investigations were severely restricted on the Apollo missions owing to conflict with the principal operational objectives. Furthering the understanding of the effects of spaceflight on human physiology therefore had to rely, almost exclusively, on comparison of pre-flight and post-flight observations. These were carefully selected to focus attention on the areas which appeared most likely to be affected; for example, cardiovascular function. Other areas were also investigated for corroborative information and unforeseen changes.

It was the output from the bioinstrumentation harness worn by the crew that was monitored by the Flight Surgeon in the MOCR and by the Aeromed flight controllers in the Life Sciences SSR.

For the Apollo flights, these included:

- Dr. M. Keith Baird.
- Dr. Charles Berry.
- Dr. Kenneth N. Beers.
- Dr. A. Duane Catterson.
- Dr. Duane Coons.
- Dr. Royce Hawkins.
- Dr. George Fred Humbert.
- Dr. Sam Pool.
- Dr. John Ziegelschmid.

Figure 7.9 Flight Surgeons Dr. Charles Berry and Dr. Duane Coons. Photo courtesy of NASA.

Data from the biotelemetry of the spacecraft were displayed at consoles at the launch and mission control centers. The consoles were manned continuously by medical personnel during a mission. Heart and respiration rates were displayed in digital form; electrocardiogram and impedance pneumogram data were presented on a cathode ray oscilloscope. See Section 5.5.3 Life Sciences SSR.

In summary, the 29 Apollo astronauts accumulated 7506 hours of spaceflight experience without encountering any major medical problems. Perhaps the most significant post-flight medical finding was the lack of any pathology attributable to spaceflight exposure. Those physiological changes which did occur were all reversible within a two-to-three-day period, with the exception of the Apollo 15 crew which required a fortnight to completely return to pre-flight baselines. The most important physiological changes seen were cardiovascular deconditioning, reduction of red blood cell mass, and musculoskeletal deterioration. Because all medical objectives of the Apollo program were successfully achieved, a sound medical basis existed for committing man to the prolonged spaceflight exposure of Skylab.

Now, a half century later, both men and women have proven that humans can spend a considerable time in space. While the record for a single flight exceeds a year, no one has spent the continuous time in space needed for a mission to Mars. Only Gennady Padalka has spent that much time in space but it was accumulated across five missions. The ISS is now the proving ground for the various types of doctors investigating the ability of humans to withstand the rigors of space travel. The NASA Human Research Program focuses on long duration flights. This has well defined research goals and objectives designed to mitigate risks. Venturing great distances into interplanetary space will pose unique medical challenges but the rewards for mankind will be substantial.

7.9 PUBLIC AFFAIRS OFFICERS

The NASA Office of Public Affairs (PAO) is responsible for media and public relations. Its mission is to "provide for the widest practicable and appropriate dissemination of information to the media and general public concerning NASA activities and results." The function was mandated by the National Aeronautics and Space Act of 1958. Services include advising the NASA Administrator on issues relating to communications as well as public and media relations; policy guidance, advice and consultation to Headquarters as well as Field Installations on public affairs issues; coordination of resources to news media and the public; maintaining communication channels to the news media and the general public. One of their most visible functions is to provide someone to explain the events occurring during all phases of a mission: launch, on-orbit, and during recovery. When somebody is assigned these duties during a mission they call themselves "commentators."

The first NASA public affairs officer for Project Mercury was Lt. Col. John Anthony "Shorty" Powers. He was dubbed "The Voice of the Astronauts" and "The Voice of Mercury Control." Most people don't know that "Shorty" was a pilot during WW-II, and after the war he flew 185 round-trip flights during the Berlin Airlift. During the Korean Conflict, he flew 55 night missions in B-26 bombers and received the Bronze Star Medal, the Air Medal, the Distinguished Flying Cross, and a combat promotion to the rank of Major.

After Korea, he was assigned to the staff of Maj. Gen. Bernard Schriever to handle public dissemination of information on the Air Force's ballistic missile program. When NASA Administrator T. Keith Glennan was seeking a public affairs officer for Project Mercury in April 1959 he arranged for Powers to be assigned to the Space Task Group.

Powers served as mission commentator for the six manned Mercury flights, adding the term "A-OK" into the American vocabulary to indicate procedures during the missions were proceeding as planned. He left NASA when Mercury ended in 1963.

Figure 7.10 John "Shorty" Powers was the Voice of Mercury Control. Photo courtesy of NASA.

Following Powers' departure, Paul Haney was promoted to Chief of Public Affairs at the new Manned Spacecraft Center in Houston. He soon took on the additional duty of providing commentary from a console on the top row of the MOCR. This was fed to broadcast television viewers. He covered multiple flights during both the Gemini and Apollo programs. He resigned after Apollo 9, over a dispute within the Astronaut Office and MSC management over public access to some private air-to-ground transmissions.

Figure 7.11 Bob Hart, Paul Haney, Al Alibrando and Terry White in the new MOCR 2 during the recovery of Gemini 4 in June 1965. Photo courtesy of NASA.

The following (in alphabetical order) are the PAO commentators assigned to the MOCR during Gemini:

- Allison M. Bond.
- Albert "Al" Chop (created the Silver Snoopy Award).
- Paul Haney.
- Bennett W. James.
- John McLeaish.

Figure 7.12 Al Chop in the new MOCR 2 during Gemini 4 simulations. Photo courtesy of NASA.

Figure 7.13 Steve Nesbitt, Shuttle commentator from 1985 to 1988. Photo courtesy of NASA.

- Milton E. Reim.
- John E. Riley.
- Saul Bert Weil.
- Terrence "Terry" White.
- L.E. Worrell.
- Edward R. Worrell.

In addition to many of the above there were the following during Apollo and Skylab:

- Walter S. Fruland.
- Douglas K. Ward.
- Don J. Green.
- Bruce J. Gordon.

There were other PAO officers in the MOCR who were not Commentators. See Appendix 2 for the Mission Manning Lists.

8

Abandoned in Place

I have taken the title of this chapter from the book published by Roland Miller in 2016. In *Abandoned in Place* he uses photography to both explore and document America's early space launch and research facilities. He employs photography to lend historical and artistic insight to the preservation and portrayal of abandoned, deactivated, and repurposed sites in a way that surpasses the official government documentation. I think the Apollo Mission Control Center is also a place which deserves preservation. This book is my way of doing that, but by using existing photography and adding both time and mission context.

8.1 TIME TO MOVE ON

By the close of the era of lunar landings in 1972, MOCR 2 was deactivated and reconfigured for the Space Shuttle. While portions of it were used for the Shuttle Approach and Landing Tests in the mid-1970s, the facility did not support space missions again until 1982. With MOCR 2 out of service, MOCR 1 picked up the load for Skylab, Apollo-Soyuz, and the Shuttle flights all the way to 1998, when STS-88 launched the first ISS component, Node 1 named Unity, to connect with the Russian Zarya module. MOCR 2 reopened in 1982 with STS-5. It saw its last flight, STS-53, in December 1992 when Discovery launched a DOD payload. In that manner MOCR 2 finished a long period of 27, on again and off again, years of support from 1965 through 1992. In the meantime, the technology, as well as the spaceflight missions, had moved on. Once again, Chaucer was right: "Time and Tide Wait for No Man."

During the 1980s, the National Park Service realized that there were a lot of facilities across NASA which had exceeded their useful life and were just being abandoned. The Johnson Space Center already knew that the space missions had changed, and the new world of ever smaller and faster computers were changing the way that flight operations

M. von Ehrenfried, *Apollo Mission Control*, Springer Praxis Books,
https://doi.org/10.1007/978-3-319-76684-3_8

could support the new missions. By the late 1980s plans were already underway to create a new control center in another building. Was Building 30 obsolete? Once it was declared a National Historic Landmark, the National Park Service defined many restrictions as to what could and could not be done with those rooms that were included in the definition of a Historic Landmark. Other rooms were not included in these restrictions. Many of those became offices and storage facilities. Some were used in ways that they really shouldn't have been used. Tours continued to show the public the old mission control center.

8.2 CHANGES IN TECHNOLOGY

It was clear that Flight Operations management was going to need a facility that was not only unencumbered by the NPS restrictions but one that would require a marked increase in technology to support the Shuttle and the International Space Station. After all, most of the technology being used was truly obsolete. In 1992, construction work began on a new five story building adjoining Building 30. By 1995 it was operational with two new control rooms, now called Flight Control Rooms instead of Mission Operations Control Rooms; phonetically, these were Fickers instead of Moukers.

Even in the decade before the center became a National Historic Landmark it was very clear that computer technology was roaring ahead of what NASA was using. Mainframes were advancing of course, but there were new smaller, faster and more capable computers. Studies were being conducted to evaluate the latest technology and the possible cost savings in terms of maintenance and operations over the existing system. The following are examples of computer advancements with reference to the ongoing space missions:

1976
This was just after Apollo-Soyuz and while the Shuttle was being designed and built. Apple I was a desktop computer designed and literally hand-built by Steve Wozniak. His friend and new business partner Steve Jobs had the bright idea of selling it. The product came complete with a main logic board, switching power supply, keyboard, case, manual, game paddles, and a cassette tape containing a game. You supplied the TV.

1977
Only Shuttle Orbiter Enterprise test flights were being made at this time. The Apple II found popularity far beyond the hobbyist community. In conjunction with a color television, the Apple II produced brilliant color graphics that were very appealing for applications. Millions of Apple IIs were sold between 1977 and 1993, making it one of the longest-lived lines of personal computers. Apple gave away thousands of Apple IIs to schools, giving a new generation their first access to personal computers.

1980
Space Shuttle Columbia was almost ready for its inaugural flight. The Motorola 68000 microprocessor had a processing speed far exceeding its contemporaries. This processor found its place in powerful work stations intended for graphics-intensive programs of the kind common in engineering.

1981–1982

MOCR 1 was supporting the first four Shuttle flights. The Commodore 64 sold for $595 with 64 kilobytes of RAM and very impressive graphics. Thousands of software titles were released over the lifespan of the C64 and by the time it was discontinued in 1993 it had sold more than 22 million units. It was recognized by the 2006 *Guinness Book of World Records* as the best-selling single computer of all time. The first IBM PC was based on a 4.77 MHz Intel 8088 microprocessor and ran the Microsoft MS-DOS operating system. This revolutionized the realm of business computing, and became the first PC to gain widespread adoption by industry.

1983–1984

MOCR 2 was routinely supporting Shuttle flights. Apple introduced an entirely new computer called the Macintosh. This was the first successful mouse-driven computer with a graphical user interface and was based on the Motorola 68000 microprocessor. Applications included MacPaint which made use of the mouse, and MacWrite which introduced WYSIWYG (What You See Is What You Get) word processing. These applications kicked off a revolution in publishing.

1984–1985

MOCR 2 was still supporting the Shuttle. On October 3, 1985, Apollo Mission Control became a National Historic Landmark. Michael Dell set up PCs Ltd in 1984. The dorm-room headquartered company sold IBM PC-compatibles built from stock components. He dropped out of school to focus on his business. In 1985 the company produced the first computer of its own design, known as the Turbo PC. By the early 1990s, Dell was one of the leading computer retailers. Meanwhile, Internet Protocols and Transmission Control Protocol and Internet Protocol (TCP/IP) received a boost when the U.S. National Science Foundation established the NSFNET that linked five supercomputer centers.

1986–1987

The loss of Challenger and the ensuing period of recovery during which NASA revised the kinds of mission that the Shuttle would accept. Compaq beat IBM to the market when it announced the Deskpro 386, the first computer to use Intel´s new 80386 chip, a 32-bit microprocessor with 275,000 transistors on each chip. At 4 million operations per second and 2 megabytes of memory, the 80386 gave PCs as much speed and power as older mainframes and also minicomputers. In addition, the new chip made graphical operating environments for IBM PC and PC-compatible computers feasible. The architecture that allowed Windows and IBM OS/2 was retained in subsequent chips. The first IBM system to use Intel´s 80386 chip shipped more than 1 million units by the end of the first year. IBM released a new operating system at the same time, OS/2, which allowed a mouse to be used with IBM PCs for the first time. Many credit the PS/2 for making the 3.5-inch floppy disk drive and video graphics array (VGA) standard equipment for IBM computers.

1989–1992

MOCR 2 was deactivated. English programmer and physicist Tim Berners-Lee submitted two proposals for what he called the "Web." By Christmas of 1990 he had prototyped the "World Wide Web" in only three months using an advanced NeXT computer.

Microsoft shipped Windows 3.0. Compatible with DOS programs, it was the first successful version of Windows and it offered good enough performance to satisfy PC users. In this version, Microsoft updated the interface and produced a design that allowed PCs to support large graphical applications for the first time. It also allowed multiple programs to run simultaneously.

Designed by Finnish university student Linus Torvalds, the Linux kernel was released to several Usenet newsgroups. Almost immediately, enthusiasts started developing it, such as adding support for peripherals and improving its stability. In February 1992 Linux became "free software" (or as its developers insisted on saying after 1998 "open source"). Linux also incorporated some elements of the GNU operating system and is used today in devices ranging from smartphones to supercomputers.

As a result of a change in policy by the National Science Foundation in 1991, the Internet was a publicly accessible network with no commercial restrictions. Four years later the Internet's backbone (main high speed lines and nodes) was turned over fully to private industry. Eight volunteers responded, and made the UNIX, Mac, and PC browsers.

1994

Shuttles were being flown out of MOCR 1, and planning was underway for the new control center. The Pentium was the fifth generation of the 'x86' series of microprocessors released by Intel. It served as the basis for the IBM PC and its clones and introduced several advances that made programs run faster; e.g. the ability to execute several instructions at the same time and specific support for graphics and music.

1995

On July 13 the Shuttle Discovery was launched and controlled out of the new $250 million mission control center. At its heart was a network of 200 generic computer workstations, each able to execute 120 million operations per second. Flight controllers now had color monitors with advanced graphic displays, and custom software which gave access to the latest data at the click of a mouse. In effect, this was "off the shelf," because the computers were what the company UNIX already had available. When better technology became available, NASA simply bought new computers. The new mission control center's consoles were individual computers, so people could work independently, unlike before when everyone was tied into the mainframe. The new consoles had far better graphic capabilities to enable controllers to participate in video conferences. In addition, flight controllers had laptops.

Also that year, Microsoft introduced Windows 95. This marked the first time that Microsoft had implemented the advanced power management specification with control in the operating system. The system also ushered in the importance of the CD-ROM drive in mobile computing, and kicked off the shift to the Intel Pentium processor as the base platform for notebooks

Figure 8.1   The new Flight Control Room, circa 1995. Photo courtesy of NASA.

Therefore the pace of technological advance had made the Apollo Mission Control Center and all almost everything in Building 30 obsolete, and the new building with the new control center picked up where it left off. It was, indeed, time to move on! Even this control center was upgraded several times over the past two-plus decades.

9

Restoration

9.1 THE VISION

The Apollo Mission Control Center is far more than just a site where history was made. It is a symbol of the greatness of the United States of America – an iconic place where people can go and see for themselves where this country actually did what it said it would do by landing men on the Moon and returning them safely to the Earth. This place, now tattered and worn, and sorely in need of costly repairs, is where dedicated teams achieved such extraordinary tasks over several decades that now prompt generations of scientists, engineers, and astronauts to tackle the technological and scientific challenges of today and tomorrow. The fully restored Apollo Mission Control Center will pay tribute to those achievements and inspire future generations to achieve their own greatness. Those of us who worked in this grand facility are comforted to know that we worked in this historic place; a place thought of by some as a "cathedral" of different kinds of spirits; a "can do" spirit, a "team" spirit, a "national" spirit and one with a backup "holy spirit" that was surely always there.

It all started with the work of a National Park Service researcher three decades ago.

9.2 THE NATIONAL PARK SERVICE

There are over 2500 National Historic Landmarks (NHL) across all fifty states, DC, all US territories and associated political states, and one on foreign soil (in Morocco). They can be either public or private property and are overseen by the National Park Service (NPS). Created in 1960, the National Historic Landmark Program is run by the Department of the Interior (DOI) to recognize and honor the nation's cultural heritage.

The Johnson Space Center has two NHLs: the Space Environment Simulation Laboratory and the Apollo Mission Control Center, the latter being the subject of this book. There are additional space related sites in Florida and near other NASA Centers.

© Springer International Publishing AG, part of Springer Nature 2018
M. von Ehrenfried, *Apollo Mission Control*, Springer Praxis Books,
https://doi.org/10.1007/978-3-319-76684-3_9

The NPS has an ongoing relationship with NASA on other projects; for example, NASA and the NPS signed a Memorandum of Agreement (MOA) for collaboration on mutually beneficial Earth science programs for the preservation, enhancement, and interpretation of U.S. natural resources. This partnership also advances NASA's mission to understand and protect our home planet and inspire the next generation of explorers.

Interest in all of the NASA space related facilities dates to a study which was started in 1981, and a report by Dr. Harry A. Butowsky in 1984 "Man in Space: Excerpts from a National Historic Landmark Theme Study (Phases I and II)" in which he identified 27 space related facilities, spacecraft, and other items, with the Apollo Mission Control Center being number 23. This led to its designation by the NPS as a National Historic Landmark in 1985, and eventually kicked off the effort to define just how the control center ought to be preserved.

9.3   COLORADO STATE UNIVERSITY

Established in 2007, the award-winning Public Lands History Center (PLHC) integrates research, education and outreach in the best tradition of a land-grant university. Projects with the National Parks and other public land management agencies enlist faculty, graduate students, and other researchers to undertake a program of historical research to directly inform current resource management challenges.

The PLHC works with federal, state and local entities as well as other partners to provide historical research services. It conducts preservation activities such as National Register and National Historic Landmark nominations and amendments, Historic Furnishings Reports, context reports, surveys, and record management. In 2013, the Program Manager, Maren Bzdek, partnered with the Partnerships & External Relations Unit of the National Park Service's Intermountain Regional Office to consult with NASA JSC on the Apollo Mission Control Center. Within that Unit was the Heritage Partnerships Program, headed by Greg Kendrick and Christine Whitacre. They arranged for the National Park Service to enter into a cooperative agreement with Colorado State University, and assigned Associate Professor of History Dr. Janet Ore as the Principal Investigator for the project.

This work involved Maren Bzdek researching the mission control center and writing a report. Her investigation included:

- Visiting Jean Grant and Lauren Meyers at the University of Houston-Clear Lake Archives, where some of the JSC History collections are archived.
- Visiting Rice University's Woodson Research Center and reviewing the Jack McCaine NASA Papers Collection.
- Visiting the National Archives in Ft. Worth, Texas, where JSC records are housed.
- Consulting with Quana Childs at the Texas Historical Commission.
- Consulting with retired National Park Service employee Kim Sikoryak with regard to the visitor interpretive plan.
- Consulting with the National Park Services' Greg Kendrick and Christine Whitacre throughout the two year process.
- Participating in the 2014 JSC project workshop.

This work resulted in the June 2015 "Historic Furnishings Report and Visitor Experience Plan: Apollo Mission Control Center National Historic Landmark," which describes the challenges and limitations of planned restoration work and provides guidance to the contractors that will actually undertake the restoration.

9.4 NASA'S JOHNSON SPACE CENTER

In August 2014 JSC employees, representatives from the National Park Service, and the Colorado State University staff and consultants held a workshop at JSC on the preservation and visitor experience of the Apollo Mission Control Center. This two day workshop was held as part of an Interagency Agreement between NASA and the NPS. It discussed the restoration of the now-designated National Historic Landmark and how it would be used for visitors once the control center was restored. Their goal was to initiate the preparation of a Historic Furnishings Report and Visitor Experience Plan that was published in June 2015.

The participants were:

- Jeannine Aquino, JSC Office of External Relations.
- Marilyn Blevins, JSC Office of Planning & Integration.
- Edward Fendell, JSC Flight Controller, Retired.
- Dennis Hehir, JSC Flight Operations Division.
- Elizabeth LeBlanc, JSC Office of External Relations.
- Charles Noel, JSC Office of Planning & Integration.
- William Owen, JSC Building 30.
- Jennifer Ross-Nazzal, JSC History Office.
- Sandra Tetley, JSC Office of Planning & Integration.
- Rebecca Wright , JSC History Office.
- Maren Bzdek, Colorado State University.
- Christine Whitacre, National Park Service.
- Greg Kendrick, National Park Service.
- Kim Sikoryak, Colorado State University, Retired.

In the summer of 2017, with funding for the project in place, a working group was established to oversee the restoration of the Apollo Mission Control Center. This JSC Apollo NHL Restoration Project Working Group included participants from JSC and Space Center Houston, as well as retired NASA flight controllers who had actually staffed the control center during their careers. This group (all JSC unless otherwise identified) included:

- Jim Thornton, Project Manager.
- Caasi Moore, Project Engineer.
- William Harris, CEO of Space Center Houston.
- Tracy Lamm, COO of Space Center Houston.
- Amy Xenofos, Legal Office.
- Thomas Morrow, Legal Office.

- Susan White, External Relations.
- Todd Pryor, Center Operations.
- Sandra Tetley, Historic Preservation Officer .
- Dr. Rebecca Klein, NASA Headquarters Federal Preservation Officer.
- Rena Schlachter, Master Planner.
- Troy LeBlanc, Chief, Mission Support Division.
- Joel Walker, Director of Center Operations.
- Brian Kelly, Director of Flight Operations.

The following former flight controllers were:

- Gene Kranz.
- Ed Fendell.
- Spencer Gardner.
- Bill Reeves.

Figure 9.1 The JSC Apollo MCC Restoration Consultation Meeting of April 2017. Sandra Tetley, the Historic Preservation Officer, is the one holding the book.

9.5   OTHER SUPPORTING ORGANIZATIONS

9.5.1   The Advisory Council on Historic Preservation (ACHP)

The ACHP is an independent federal agency which promotes the economic, educational, environmental, sustainability and cultural values that arise from historic preservation. It also advises the President and Congress on national historic preservation policy.

On September 6, 2017, the ACHP announced its partnership in restoring the Apollo Mission Control Center. This was in response to a letter that Gene Kranz wrote to ACHP Chairman Milford Wayne Donaldson on January 26 of that year pointing out the condition of the Apollo MCC. See Appendix 1 for a copy of the letter.

NASA is not authorized by law to accept private contributions dedicated to specific projects such as this restoration, so private funding was essential, and there had to be a funding mechanism that provided appropriate accountability.

Fortunately the ACHP is authorized to administer donations using the 1976 amendments to the National Historic Preservation Act which President Lyndon Johnson signed into law in 1966. In 2003 President George W. Bush signed an Executive Order that directed the ACHP to use its donation authority to assist other federal agencies in the preservation of historic properties. Support of the Apollo MCC funding is the first use of this authority by the ACHP.

After the Apollo Mission Control Center was designated a National Historic Landmark in 1985, NASA, the ACHP, and the National Conference of State Historic Preservation Officers met to discuss the activities that would be carried out and to identify specific stipulations that NASA had to ensure were executed. Annual coordination meetings were defined for NASA to brief the other parties on progress. This meeting not only covered the Apollo Mission Control Center, but a total of twenty NASA National Historic Landmarks all across the country. The agreement was signed in September 1989.

NASA continued to fly Shuttles out of the control center through STS-53 in December 1992, then it was simply abandoned.

After years of visitors, unabridged access, and declining budgets, the state of the control center declined to the point that in September 2015 the National Park Service listed it as a "Threatened" National Landmark. In response, JSC took the actions as defined above in Section 9.2 and also met with the ACHP to formulate the agreement shown in Figure 9.2.

9.5.2 The National Trust for Historic Preservation (NTHP)

The NTHP is a privately funded nonprofit organization whose mission is to save historic sites that represent America's diverse cultural experience. In doing so, it takes direct action and inspires broad public support.

9.5.3 The Texas Historical Commission (THC)

The THC is the state agency for historic preservation. Their mission is to protect and preserve the state's historic and prehistoric resources for the use, education, enjoyment, and economic benefit of present and future generations. It serves as the State Historic Preservation Office, in accordance with the National Historic Preservation Act of 1966 as amended. They monitor the economic impact of all historic preservation activities in Texas. The Apollo Mission Control Center is just one of many sites that fall under their purview.

For Immediate Release:
September 6, 2017

Contact: Matt Spangler
(202) 517-1481
mspangler@achp.gov

Megan Sumner
(281) 483-5111
megan.c.sumner@nasa.gov

ACHP ANNOUNCES PARTNERSHIP TO
RESTORE APOLLO MISSION CONTROL CENTER

HOUSTON – Restoration of NASA's Apollo Mission Control Center (MCC) will get underway this fall, thanks to a $5 million fundraising campaign spearheaded by the nonprofit Space Center Houston, which is partnering with the Advisory Council on Historic Preservation (ACHP) to administer the funds.

Since being decommissioned in the 1990s, the Apollo MCC–a National Historic Landmark housed within the Johnson Space Center (JSC) in Houston–was assigned "threatened" status by the National Park Service in 2015. The predicament led Gene Kranz, flight director for the famed Apollo and Gemini missions in the 1960s, to ask ACHP Chairman Milford Wayne Donaldson for the ACHP's assistance in restoring the Apollo MCC.

NASA and its partners aim to renovate the Apollo MCC to its late 1960s state in time for the celebration of the 50th anniversary of the first moon landing in July 2019. As part of the work, the Mission Control consoles will be reanimated to represent the Apollo 15 mission configuration and adorned with an array of personal items and replicas of authentic documents to recreate the historic scene setting. The restoration effort, which affects multiple spaces within the Apollo MCC, will be conducted in accordance with the Secretary of the Interior's Standards for the Treatment of Historic Properties and coordinated with the Texas Historical Commission, the state authority for historic preservation.

The space agency is not currently authorized by law to accept private contributions dedicated to specific projects, such as the Apollo MCC restoration. Thus the private funding has been secured by Space Center Houston and will be administered by the ACHP. The funding will be phased in over a period of 23 months. The nonprofit Space Center Houston is a leading science and space learning center and official visitor center for the JSC. The ongoing campaign has raised more than $4 million, including $3.5 million from the City of Webster, Texas.

The ACHP is authorized to administer the donation through 1976 amendments to the National Historic Preservation Act, which was signed into law by President Lyndon Johnson in 1966. An Executive Order signed in 2003 by President George W. Bush directs the ACHP to use its donation authority to assist other federal agencies in the preservation of historic properties. The NASA award is the first such use of the authority to date.

"The donation for work at the Johnson Space Center represents an excellent example of how public-private partnerships, such as that between the City of Webster, Space Center Houston, NASA, and the ACHP, can help preserve some of the sites most important to telling our nation's story," Donaldson said. "We hope other federal agencies will explore these types of partnerships for properties in their inventories that are in need of saving."

ADVISORY COUNCIL ON HISTORIC PRESERVATION

401 F Street NW, Suite 308 • Washington, DC 20001-2637
Phone: 202-517-0200 • Fax: 202-517-6381 • achp@achp.gov • www.achp.gov

Figure 9.2   The Partnership formed by NASA and ACHP. Document by ACHP press release.

9.5.4 Manned Spaceflight Operations Association (MSOA)

The MSOA was organized in October 2017, with the mission to perpetuate the memory of those who came forward from all over America to fulfill President Kennedy's challenge to send men to the Moon. It recognizes and honors those who planned, trained and supported the many missions flown out of the Apollo Mission Control Center.

The association supports the preservation and maintenance of the MCC, and the restoration of the consoles, displays and support equipment. It also promotes the communication of the history of the control center and the benefits of space exploration to society. It intends to use this history to inspire new generations to educate and prepare themselves for future space challenges. It communicates to those who have supported spaceflight and to members of the public via website www.mannedspaceops.org.

The Board of Directors of this organization are (alphabetically):

- Spencer H. Gardner
- Gerald D. Griffin
- Fred W. Haise
- Jeffrey M. Hanley
- Frank E. Hughes
- Eugene F. Kranz
- Glynn S. Lunney
- William D. Reeves
- Manfred von Ehrenfried.

The current officers are:

- William D. Reeves, President
- Linda Ham, Vice President
- Spencer H. Gardner, Secretary
- Jan Weede, Assistant Secretary
- David W. Whittle, Treasurer
- Arthur L. Schmitt, Assistant Treasurer.

9.6 FUNDING THE PROJECT

9.6.1 The Webster Challenge

William Michael Rodgers, CPA, Director of Finance and Administration of the City of Webster, Texas has provided an account of how the funding required to restore the Apollo Mission Control Center was acquired:

The Houston Chronicle published a few articles in late 2016 on the need to restore the Mission Operations Control Room at Johnson Space Center. Plans had been made; however, the $3.1 million cost of the project was overwhelming. The budget for NASA did not include funding for projects like this, so the project lay in limbo. After reading these articles, I told my City Manager, Wayne Sabo, that I had a crazy idea to propose. The restoration project simply had to be done. By sponsoring the project, the City could recognize the amazing accomplishments of the people who lived and worked in Webster during the Apollo era. We would also garner international exposure during the process. Instead of running me out of his office, Wayne actually liked the idea.

In December 2016, Wayne and I met with William T. Harris, President and CEO of Space Center Houston, and Kim Parker, Vice President of Development at Space Center Houston to discuss the details of the restoration project. William discussed the work that was planned and its $3.1 million price tag. Wayne and I looked them both in the eyes and asked, "What would you think if we paid for this entire project?" Not only could we fund the restoration, Webster could offer an additional $400,000 as a dollar-for-dollar match towards future maintenance of the facility. Of course, the Webster City Council would have the final say on whether the sponsorship would occur. After a few moments of stunned silence, William and Kim jumped at the opportunity.

The City of Webster has been the tourism marketing partner with Space Center Houston, the Official Visitor Center for NASA Johnson Space Center, for the past seventeen years. Thousands of the visitors to Space Center Houston and JSC stay in one of the eighteen hotels located in Webster. It is the reason why the City of Webster had amassed nearly $5 million in hotel occupancy taxes, enough to pay for the project. The hotel occupancy tax is a special type of tax that is collected by hoteliers. State law restricts the way a municipality can use these funds to specific things, including "historical restoration and preservation projects that would be frequented by tourists." Funding the restoration project clearly falls within the parameters set by law.

As Webster's mayor, Donna Rogers, says: "The city of Webster, its hoteliers and hospitality partners are dedicated to sustaining and enhancing the community. This donation is one way to provide exceptional learning opportunities that attract people from around the world."

In April 2017, the City Council of the City of Webster approved a Gift Agreement with Space Center Houston that sets forth the parameters for the City's contribution of hotel occupancy taxes in an amount not to exceed $3,500,000. As reflected in the Gift Agreement, Space Center Houston will provide permanent recognition within the lobby area of Building 30 on the Johnson Space Center campus; promote Webster hotels on the Space Center Houston website and other digital media throughout the "Webster Challenge" and create an exhibit or video that describes the connection between NASA and the City of Webster.

The Webster Challenge crowdfunding campaign was initiated soon afterwards. Over 4,200 backers contributed more than $500,000 towards the project. This has been a great success for everyone involved. The City of Webster will complete its sponsorship funding in September, 2018.

Figure 9.3 The City of Webster "Supporters" Wayne J. Sabo and Michael Rodgers. Photo courtesy of Treasurer Mike Rodgers.

9.6.2 Space Center Houston

As a project of the Manned Space Flight Education Foundation, Space Center Houston is the nonprofit 501(c)(3) Official Visitor Center of JSC. It relies on private contributions and ticket revenue to fund its operations. Projects like the restoration of the Apollo Mission Control Center are reliant entirely on private contributions.

With the 50th anniversary of Apollo 11 looming in 2019, restoration of the Apollo Mission Control Center became urgent. Retired NASA flight operations team members began working with Space Center Houston to secure the funds to restore the site and create a world-class visitor experience that will inspire future generations through this amazing story of technological and human achievement.

In 2017, Space Center Houston launched a $5 million campaign to raise funds to help support a major restoration of the MOCR. $3.5 million had already been raised through a generous lead gift from the City of Webster, Texas. For further information go to www. spacecenter.org.

9.6.3 Kickstarter

From July 20 (the Apollo 11 landing anniversary) through August 19, 2017, Space Center Houston conducted a crowdfunding effort called "The Webster Challenge: Restore Historic Mission Control" on Kickstarter, the fund raising platform of the American Public Benefit

Corporation. The Webster Challenge invited people from around the world to donate over a 30-day period to raise funds for the restoration. A total of 4,251 people pledged $506,905, of which $400,000 was generously matched by the City of Webster. This permitted the restoration of the Apollo Mission Control Center to get underway.

Figure 9.4 Kickstarter Flight Operations Supporters. This photo was on the Space Center Houston website to encourage people to contribute to the fund raising effort. From front to back are Glynn Lunney, Milt Windler, Gene Kranz, Ed Fendell, Bill Moon, and Bob Grilli. Back row L-R: Denny Holt, Gerald Griffith, Spencer Gardner, and Bob Nute. Photo courtesy of Space Center Houston.

9.6.4   Gift Agreement

The following is the formal gift agreement between the City of Webster and the Manned Space Flight Education Foundation doing business as the Space Center Houston.

GIFT AGREEMENT

This Gift Agreement ("Agreement"), effective as of April 4, 2017 ("Effective Date"), is made and entered into by and between The City of Webster, whose address is 101 Pennsylvania Ave, Webster, TX 77598 ("Donor") and the Board of Directors of the Manned Space Flight Education Foundation, Inc. dba Space Center Houston for the purpose of funding the historic restoration of the Apollo Mission Operations Control Room and related areas ("Restoration Project") at NASA's Johnson Space Center. Based upon the Recitals below, and in consideration of the mutual promises and benefits hereunder, the parties agree as follows:

RECITALS

Donor wishes to make a charitable gift to Space Center Houston for the purpose of funding historical and preservation projects that encourage tourists to visit preserved historic sites in its vicinity and promoting tourism and the convention and hotel industry. Accordingly, Donor wishes to provide funding for the Restoration Project that consists of the Mission Operations Control Room and related areas associated with the 1969 Apollo Moon Landing, located in Building 30 on the campus of NASA's Johnson Space Center, as set forth in this Agreement. Space Center Houston wishes to conduct a global crowdfunding campaign (such as "Kickstarter") to raise awareness of the project and encourage broad public participation in the fundraising effort.

Space Center Houston desires to accept such gift, subject to the terms and conditions set forth in this Agreement.

AGREEMENT

1. **Gift.** Donor hereby pledges to Space Center Houston the following gift: $3,500,000 ("Gift").
2. **Payment of the Gift.** The Gift is an irrevocable pledge that will be paid to Space Center Houston over a period of two years. Payments in support of this pledge will begin immediately upon the execution of this Agreement with an initial payment of $1,800,000 and will continue thereafter according to the following schedule based on the project construction and estimated dates:

Amount of payment by Donor	Due Date
$1,000,000	December 1, 2017
$300,000	April 1, 2018
$400,000	September 30, 2018 (Kickstarter match, upon receipt of donation records)

Space Center Houston will provide status reports on the renovation project one week prior to each payment due date. Upon timely submission of invoices (and donation receipts for Kickstarter), payments shall be paid by Donor to Space Center Houston via check or electronic funds transfer. Donor may accelerate the payment of any or all of this pledge at any time in Donor's discretion so long as the cumulative total of all gift payments meets the foregoing schedule.

3. **Use of the Gift.** The Gift shall be restricted for the following purposes:

$3,100,000 is designated for the Mission Operations Control Historic Restoration project.

$400,000 is designated for a challenge grant to be used in Space Center Houston's "Kickstarter Campaign" as the "Webster Challenge" to complete the funding of the historic restoration project. The grant will match all contributions to the Webster Challenge dollar for dollar up to an amount not to exceed $400,000.

4. **Acknowledgment.** In consideration for the Gift, Space Center Houston will acknowledge the Gift as follows:

A plaque dedicated to the City of Webster, to include names of all Council members in office in February 2017 and members in office at the time the project is completed and inaugurated, in the lobby area of Building 30 on the Johnson Space Center campus. Subject to the terms of this Agreement, the recognition plaque will last for the useful life of the Facility.

A News Release on or about the date the gift is confirmed and a private gift announcement event for the Webster City Council and Webster hoteliers on a mutually agreed upon date.

In digital communications, including social media, announcing the gift, the "Webster Challenge" and later giving updates. The communications will acknowledge that the city is using Hotel Occupancy Tax funding for the gift.

Prominent recognition as part of the Kickstarter Campaign will be given on social media.

Quarterly public reports on the status of restoration project and status of the Kickstarter Campaign, including the "Webster Challenge." The communications will acknowledge that the city is using Hotel Occupancy Tax funding for the gift.

Acknowledgement of the gift, including that the city is using Hotel Occupancy Tax funding, on the Space Center Houston webpage that describes the Historic Mission Control restoration project and on the City of Webster's Preferred Partner webpage.

An exhibit or video, developed in collaboration with the City of Webster that describes the connections between NASA, the City of Webster, Webster hospitality partners and the local Bay Area community.

A Dedication Ceremony on or about the time the Historic Mission Control Restoration is completed in 2019.

5. **Termination.** In addition to any rights and remedies available at law, SCH may terminate this Agreement and all rights and benefits of the Donor hereunder in the event of any default in payment of the Gift as provided in this Agreement.

6. **Modification of Donor Recognition.** If during the useful life of the MOCR and related areas, it is closed, deconstructed, destroyed or severely damaged, significantly renovated, upgraded, or modified; relocated, or replaced, then the donor recognition will cease. In such event, however, the Donor, if available, and in consultation with and as mutually agreed by SCH and JSC, will have the right, for no additional payment, to have another available and equivalent area for donor recognition.

7. **Publicity.** For purposes of publicizing the Gift, Space Center Houston will have the right, without charge, to photograph representatives of the Donor and use the names, likenesses, and images of the representatives of the Donor in photographic, audio-visual, digital or any other form of medium (the "Media Materials") and to use, reproduce, distribute, exhibit, and publish the Media Materials in any manner in whole or in part, including in brochures, website postings, informational and marketing materials, and reports and publications describing Space Center Houston's development and business activities.

8. **Assignment.** This Agreement and the rights and benefits hereunder may not be assigned by either party without the prior written consent of the other party, which consent shall be in the sole and absolute discretion of the non-assigning party.

9. **Entire Agreement.** This Agreement constitutes the entire agreement of the parties with regard to the matters referred to herein, and supersedes all prior oral and written agreement, if any, of the parties in respect hereto. This Agreement may not be modified or amended except by written agreement executed by both parties hereto. The captions inserted in this Agreement are for convenience only and in no way define,

limit, or otherwise describe the scope or intent of this Agreement, or any provision hereof, or in any way affect the interpretation of this Agreement.

10. **Governing Law and Venue**. This Agreement will be governed by and construed in accordance with the laws of the State of Texas without regard to any conflict of laws rule or principle that might refer the governance or construction of this Agreement to the laws of another jurisdiction. Subject to the sovereign immunity of the State of Texas, any legal proceeding brought in connection with disputes relating to or arising out of this Agreement will be filed and heard in Harris County, Texas, and each party waives any objection that it might raise to such venue and any right it may have to claim that such venue is inconvenient.

11. **Independent Contractor**. The parties shall at all times act independently. Nothing contained in this Agreement shall be construed to make either party the partner, joint venturer, principal, agent or employee of the other party hereto.

12. **Force Majeure**. Neither party shall be held responsible for any delay or failure in performance of any part of this Agreement to the extent that such delay or failure is caused by fire, flood, explosion, war, strike, embargo, government requirement, civil or military authorities, Act of God or by the public enemy, or other causes beyond the control of Donor or SCH. If any force majeure condition occurs, the party delayed or unable to perform shall give immediate notice to the other party and the party affected by the other's inability to perform may elect to: (a) terminate this Agreement or any part thereof as to future performances; or (b) resume performance under this Agreement once the force majeure condition ceases with an option in the affected party to extend the period of this Agreement up to the length of time the force majeure condition endured, unless written notice is given within thirty (30) days after such affected party is notified of the force majeure condition, then (b) shall be deemed selected.

13. **Board of Directors Approval**. This Agreement and the recognition provided for herein are subject to the approval by the board of the Manned Space Flight Education Foundation and this Agreement will not be effective unless and until approved by the board and the Donor.

ACCEPTED AND AGREED TO:

CITY OF WEBSTER

BY:

Name: Donna Rogers
Title: Mayor, City of Webster

MANNED SPACE FLIGHT EDUCATION FOUNDATION

By: _____ 3/27/17

Name: William T. Harris
Title: President and Chief Executive Officer

9.7   PROPOSED WORK

The restoration will return the Apollo Mission Control Center to its authentic appearance during the era of the Apollo lunar landings. To do so, a number of changes to the configuration of the consoles will be required. When the room served as mission control for the Shuttle during the 1980s those consoles were reconfigured and their technology upgraded.

The project will therefore restore each of the consoles to mimic their operational configuration at the time of the Apollo 15 mission in 1971, which was considered representative of the apex of technology in the Apollo era. This will also apply to the console arrays such as panels, switches, indicators and monitors. Visitors will therefore experience the room just as if that mission was still on the Moon.

Consoles were to be sent to the Cosmosphere in Hutchinson, Kansas. There experts in the SpaceWorks Division would restore and reanimate each console. In addition to installing representative buttons and sequences, SpaceWorks will illuminate the monochromatic displays on the cathode ray tube monitors. This reanimation of the consoles is a vital component of bringing MOCR 2 "back to life" for visitors. The large group displays on the west wall of the room will also be reactivated using appropriate projection technology to recreate the screens of the Apollo era. Lighting will be replaced in the room with dimmable LED lights, recreating the lighting settings that maximized control operations during Apollo missions. This will avoid long-term exposure to unfiltered ultraviolet damaging furnishings. Indeed, the historic "blue" overtone from the world map screen will also be recreated.

Other important details include restoring and replacing the furnishings of that era. Extensive research, including interviews with Apollo flight controllers in the MOCR 2 will confirm the activities that took place at each console, and discover what types of personal items would have been present on each console. The plan is to replicate items such as ash trays, binders, pencils, headsets and coffee cups.

The project planned to start in August 2017, but was delayed for funding and letting of contracts until January 2018. First was the restoration and reanimation of the consoles because this aspect of the project had the longest lead time. In the meantime, crews made a start on cleaning, repairing, and refurbishing the site – replacing old carpet and acquiring all the historic furnishings for a fully authentic Historic Mission Control. The anticipated completion date for the entire project is January 2019, in good time to permit visitors to celebrate the 50th anniversary of Apollo 11.

The restoration work is organized into the following areas (summarized):

- Project Management

    1. Quality Control Plan.
    2. Site Safety and Health Plan.
    3. Safety Data Sheets.

- Restoration Design

    1. Installation Design Documentation.
    2. Electrical, Lighting, Audio Visual, Consoles.
    3. Installation Documents and As-Built Drawings.

- Artifacts

    1. Curation of Historic Artifacts & Travel.
    2. Remove/Curate Artifacts from Walls & Travel.
    3. Restoration of Building 30 Dioramas.
    4. Artifact Sourcing, Recreation and Coordination.
    5. Manuals/Documentation Research Reproduction.

- Console Reconfiguration/Restoration

There are four groups/rows of Consoles, and each group will require the following work:

1. Shipping Skid fabrication.
2. Ship Skids and tools to JSC.
3. Load Consoles at JSC.
4. Ship Consoles to Cosmosphere.
5. Receive Consoles at Cosmosphere.
6. Console and Hardwar ID and Marking.
7. Console Restoration & Re-animation.
8. Pack Console Hardware.
9. Prep Consoles and Hardware for Shipping.
10. Ship Consoles and Hardware to JSC.
11. Receive Consoles and Hardware at JSC.
12. Install Consoles in the MOCR.
13. Staging.

- Seating Restoration

There are five rows in the Visitor Viewing Room, and each row will require the following work:

1. Remove seat.
2. Restore.
3. Reinstall seat.

- Restore the Viewing Room Consoles and Booth

1. Remove console and seating.
2. Restore console and seating.

- Furnishings Restoration

1. Viewing Area Desk.
2. Clean and Oil Wood Surfaces.
3. ADA Ramp and Handrail.
4. Mission Boxes and Controller Book Shelves.
5. Final Cleaning after completion of Restoration.
6. Clean A/C Grills.
7. Acquisition and Placement of Furniture.

- Floor/Carpet Cleaning/Replacement

1. Carpet Cleaning in the Viewing Room.
2. Carpet, Base and Raised Flooring.
3. Stanchion.

- Wall and Window Treatments

1. Remove Artifacts and take to storage.
2. Painting and Wallpaper.
3. Glass.

      4. Metal Cleaning, Provide/Install Doors & Frames.
      5. Door/Covering for lighting control area.

- Ceiling Restoration

      1. Cleaning.
      2. Restoration.

- Lighting

      1. Cleaning.
      2. Dimmable Lighting and Control.

- Audio/Video Equipment/Installation

      1. Local Area Network.
      2. Summary Display-Group Display Clocks.
      3. AV Equipment and Controls.

- Bat Cave Preservation

      1. Display Terminal Area Floor Cleaning.
      2. Static Equipment Replacement.

- Maintenance Plan

      1. Develop and document.

- Miscellaneous

      1. Conservation/Restoration of Gemini, Apollo and Shuttle Plaques.
      2. Conservation of Director Retirement Proclamations.
      3. Apollo Plot Board Restoration.

Figure 9.5 Tagging, labeling and recording items for shipment. Photo courtesy of collectSPACE.

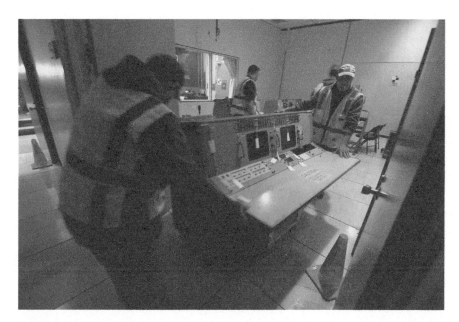

Figure 9.6 The Simulation Control Room consoles were the first to go. Photo courtesy of collectSPACE.

Figure 9.7 The first consoles to go to Cosmosphere included the Guidance Officer (GUIDO), Retrofire Officer (RETRO), Flight Dynamics Officer (FIDO), and Capsule Communicator (CapCom). Photo courtesy of collectSPACE.

Figure 9.8 Shipping the first consoles to Cosmosphere. Photo courtesy of collectSPACE.

Figure 9.9 The Spaceworks crew: Dale Capps, Don Aich, and Jack Graber. Photo courtesy of the Cosmosphere, Hutchinson, Kansas.

9.8 THE RESTORATION CONTRACTORS

The prime contractor for the work is Ayuda Companies of Denver, Colorado, an Economically Disadvantaged Women-Owned Small Business, environmentally oriented company. The President is Sonja Yungeberg and the Project Manager is Jennifer Keys. The initial contract was let on January 9, 2018 for Task 1 of the effort. The contract is managed by NASA JSC Contracting Officer Frances Davidson and Contract Specialist Michael McIntosh. The Contracting Officer's Representative Sandra J. Tetley manages the work. She is assisted by an Ayuda Historic Preservation Specialist. The JSC Construction Project Manager is Todd Pryor.

There are several subcontractors, as follows:

- Stern and Bucek Architects, Houston, Texas. The lead person is David Bucek with support from Jordan Shelton and Delaney Harris-Finch for restoration of furnishings, carpets, wall paper, ceiling tiles, and Visitor Viewing Room seating and upholstery. Supporting Stern and Bucek are Melanie Sanford from Textile Preservation Services of Texas and Jhonny Langer from Source Historical Services.
- GRAVitate, LLC, Ft. Worth, Texas. Dr. Adam C. Graves is the lead preservationist for the restoration of the Apollo Mission Control Room.
- Audio Video Guys, Inc., Houston, Texas. The lead person is George L. Weisinger providing Local Area Network, Audio/Video Systems and installation in the MOCR and surrounding areas.
- Basic Builders, Inc., Houston, Texas. The lead person is Don Baer for interior restoration work.
- Cosmosphere SpaceWorks, Hutchinson, Kansas. The President is Jim Remar, and support is provided by Jack Graber who is managing the work to restore the consoles.

Cosmosphere is renowned for their restoration of flown U.S. spacecraft for display in museums and exhibits across the globe, including artifacts that are part of the collection of the Smithsonian Institution National Air and Space Museum. Two examples of their work are the Apollo 13 Command Module, Odyssey, and Gus Grissom's Mercury capsule, Liberty Bell 7, following its recovery from the floor of the Atlantic, both of which are now on display at Cosmosphere.

In July 2017 Cosmosphere interviewed former flight controllers seated at their consoles to acquire as much information as possible in preparation for restoration work, and also solicited personal items that might be displayed in the final control center configuration. On January 25, 2018 Cosmosphere workers began to remove the consoles from the Simulation Control Room and also the first two rows of the MOCR 2, labelling every item and logging it in a database. It was the first step in the process to restore the original Mission Operations Control Room 2, the Visitor Viewing Room, and the Simulation Control Room back to the way they appeared in July 1969. (Though the overall control room is being returned to its Apollo 11 appearance, the consoles will be configured to illustrate Apollo 15 in July 1971.) On arrival at the SpaceWorks facility, conservators will carefully take apart the more than half-century old consoles in order to bring them back to life. After the first batch of consoles are returned to NASA in October 2018, the second batch, which will include the flight director's console, will be similarly processed.

In the meantime in Houston, work to restore the rooms themselves has begun, with details like the carpet, ceiling tiles, wallpaper and upholstery being cleaned or replaced to match their 1969–1971 appearance.

10

The 50th Anniversary of the First Lunar Landing

10.1 THE FINAL GOAL OF THE RESTORATION

July 16, 2019 will be the 50th Anniversary of the launch of Apollo 11, and four days later will be the anniversary of the first lunar landing. More important will be July 24 because it will celebrate the crew's safe return "before this decade is out."

It is the intent of everyone involved in the restoration effort that the Apollo Mission Control Center will be available for a ceremony to commemorate this historic national – indeed "human" – achievement. Although it isn't feasible to restore the entire Mission Control Center (Building 30), the part which is being restored includes the room that, more than any, represents the national effort to land man on the Moon and return him safely to the Earth. This is the third floor Mission Operations Control Center, known in NASA jargon as MOCR 2. If all goes to plan, the restoration will include personal and utilitarian objects which were used by the flight controllers. This attention to detail, which in the movie-making world is known as "dressing the set," should evoke powerful emotional and intellectual responses from visitors, showing that "history happened here."

Because most visitors to the building will be escorted to the Visitor Viewing Room, this is being restored too when funds become available. Of course, the first things that visitors will see through the wall of glass overlooking the MOCR 2 will be the rows of consoles and the "big screen" on the wall beyond. These will all be restored to appear as they did half a century ago. Visitors will have partial views into other rooms, in particular the Simulation Control Room, but (at least initially) will see only a large picture of the Recovery Operations Control Room affixed to its window. These rooms will be restored to a lesser degree of fidelity if and when funds become available.

10.2   THE VISITOR EXPERIENCE

The restoration team, including all those who worked on the Historic Furnishings Report and Visitor Experience Plan, are in agreement concerning why the Apollo Mission Control Center should be restored. They laid out their views as to how to best present the control center to the thousands of visitors that would surely come each year in the future. The Plan recommended the following interpretive themes that provide the foundation for the development of visitor experiences:

- NASA and the Apollo missions powerfully exemplify the human drive to explore new frontiers.
- Apollo met the challenge laid down by President John F. Kennedy in 1961 and, despite adversity and tragedy, safely landed Americans on the Moon within the decade as specified, inspiring the next generation of scientists, engineers and astronauts.
- The Apollo program demonstrated that with the commitment and support of the American people, a young, dedicated and enthusiastic engineering team could accomplish what was initially considered impossible.
- The creativity and inventiveness of the Apollo team produced enormous advances in a wide range of technologies and sciences that not only took us to the Moon but changed our lives forever.

To provide the best visitor experience, it is planned that the tours be limited to the 74 seats available in the Visitor Viewing Room. In there, visitors will be able to view live feeds on the "big screen" group displays, hear recorded Apollo crews during a lunar landing or excursion, and hear the flight controllers' conversations. These tours will be managed by Space Center Houston, and might include retired NASA flight controllers and engineers who will offer insights into what occurred in the control room and then engage in Q&A sessions.

It is also envisioned that VIP tours will be conducted with smaller numbers of visitors who will be permitted to access the MOCR itself. Of course, they will be closely controlled to avoid any damage or impact to the recently restored facility; for example, no one will be permitted to sit on a console, or to handle the objects that are on and around the consoles.

As former NASA flight controller Ed Fendell put it, "Apollo Mission Control should be restored to a degree of accuracy that will feel to visitors like the day we walked out."

10.3   WHAT IT MEANS TO THE NATION AND THE WORLD

Authors, poets, philosophers, politicians, historians, and even astronauts have all attempted to describe the "meaning" of Apollo in prose and poetry. Libraries are filled with scholarly tomes, some of which are highly technical. This book takes a different tack. It is almost a "tourist guide book" that sets out to fully describe the building and room where people executed the detailed plans designed to achieve the goals of Apollo. It is hoped the National

Historic Landmark called the Apollo Mission Control Center will bring a tangible meaning to the word "Apollo."

What does this iconic place mean to the nation, and the indeed world? It may be just a few rooms, but it is where momentous events happened. To some of the current generation it will be only a place to visit while on vacation. That 10 year old boy who watched the first lunar landing on a black-and-white TV in 1969 is around 60 now. He may be taking his grandson to Space Center Houston to see what the space program was all about when he was a kid. Part of his tour will be Building 30. Over half of the U.S. population today was not born in 1969! What do they know about the history of the space program, let alone about the Apollo Mission Control Center? Will the history of this iconic place be recorded only in the archives? By restoring it to its former glory, it will be possible for visitors to gain a visual and visceral sense of what happened there, over half a century ago.

The young people who will visit the center will be reminded that the average age in Mission Control during Apollo 11 was 28. During Project Mercury it was even younger. Apollo 11 caused a surge in college enrollment for engineers and scientists, in particular physicists. The wish to be part of as grand a venture as a flight to the Moon inspired tens of thousands of people, young and old. Today's educational programs are geared toward Science, Technology, Engineering and Mathematics (known as STEM) and are coupled with industry and government programs to further assist the students. The education level of students who are interested in careers in spaceflight is much higher now than ever. How can one measure the inspiration that the restored Apollo Mission Control Center will give to the youngsters of tomorrow? I like to think that the word "inspiration" has in its roots the word "spirit," and that a divine guidance exerts influence on the one so inspired.

Most visitors will hardly be aware that Apollo space technology spinoffs had an influence in the planning and execution of their trip. They use smart phones, computers, and the internet. Their cars are now heavily computerized. And they probably used the GPS systems to locate their hotel. No such technologies were available to plan and support the Apollo missions, nor indeed to assist the flight controllers in the Apollo Mission Control Center. With these technologies (and others) people all around the globe are connected in ways that could never have been foreseen half a century ago. In fact, their influence may be more profound than we currently realize.

For those who worked in the Apollo Mission Control Center, there are some special meanings and memories. Here are a few:

- "I believe the Apollo program will go down in history as the most significant technological achievement of the 20th century. For sure, that achievement was anchored in the development of aviation, but to leave this planet and land humans on another body in our solar system and return them to Earth was a uniquely difficult and bold accomplishment in the history of the USA. The restoration and preservation of this one-of-a-kind site is in keeping with the care taken to preserve other significant points of interest in the nation's history. The restored MOCR will take its place as a symbol of American know-how and accomplishments, and perhaps more importantly, for future generations it will preserve and tell the story about the first steps of the human exploration of the universe." – Gerry Griffin, Apollo Flight Director, Director of the Johnson Space Center.

- "I represent "the children of Apollo" who were inspired by the audacity of a nation to act so boldly as to reach for another world. In 1969 I was 8 years old, an impressionable age when we forge lifelong heroes. I was hooked, and I came to know the names of Chris Kraft and Gene Kranz as well as any astronaut's name.

  For my generation of flight controllers, the Mission Operations Wing of the original control center complex was already a "shrine" when many of us arrived at JSC in the early 1980s. Already then, the control room was a hallowed place of achievement, of excellence, of ingenuity, and of grit. We clamored to be selected to train there, and we were schooled in the rigors of flight control by the Apollo generation just before they were to move on to other challenges.

  Restoration of the room is really a small piece of the story for me. Yes it captured a moment in time when I must have wondered, "What was it like?" But the physical reality of the room is small potatoes compared to the people, the giants, on whose shoulders I have directly stood in my own career. I have met many but I would say not most, I have received the counsel of a few, and I have done my level best to emulate one or two along the way.

  And maybe therein lies some of the meaning for the nation and the world. The room represents American grit and ingenuity at its best – when the mission goes well, and when it goes very wrong. For all the "acceptable reasons" as coined by Mike Griffin (Administrator of NASA, 2005–2009) for Kennedy's challenge, the "real reasons" lay closer to our humanity. What was achieved in that room by the generation of engineers, managers and scientists was no mere stunt but the transformation of a culture to see our human condition differently, to see the Earth differently, and radically alter our concept of the possible. The room is just a room, but the consequences of what people did there echo through time in ways we have yet to perceive fully, almost half a century hence. This is my wish for the restoration. That it not be treated as just an "artifact of a bygone era" but as a symbol of the American heart and the American soul." – Jeff Hanley, Flight Controller, Chief, Flight Director's Office, and also Manager, Constellation Program.

- "As we approach the 50th anniversary of the historic Apollo 11 first landing of humans on the Moon, we must reflect on what that event meant to the nation and the world and the importance of restoring the Historic Mission Control Center in Houston, Texas that supported those operations. I have given this a lot of thought lately and draw parallels to the first voyages of the daring explorers that set sail from Europe across the unknown ocean for the sake of discovery. One can draw parallels to many historic explorations such as Lewis and Clark, Admiral Byrd, and Charles Lindberg to name a few. These explorers were blazing a trail that opened the door to new technologies and new opportunities for mankind. I see little difference between those adventures and the challenges of getting to the Moon and back. The early explorers did not have the advantage of prior knowledge of what lay ahead nor did they have the support of thousands of people to help them. They were on their own but that does not take away in any respect the dangers and risks the crews of the space program had to accept. And it was in the preparation where the needed technologies were developed that benefitted mankind for decades afterwards.

One has to remember that when President John Kennedy made the famous speech about going to the Moon and back in 1961, only a few humans had ever flown into space, none of the Centers really existed as they are today, the vehicle designs did not exist, the techniques and operations for getting there did not exist. Yet eight years later the goal was accomplished which included the building of the entire infrastructure of the space program. The Historic Mission Control is a symbol of all those accomplishments." – William D. Reeves. Flight Controller supporting the Apollo 11 Lunar Module Electrical Power System.

• "Former, and now much older flight controllers can see the obvious impact of the space program on society and the world. They came from a slide rule world with pencils; not even calculators. They had no personal computers, no smart phones, no World Wide Web, no 24/7 information flow from all around the world, no 200 channels of TV, and no digital this and that. One obvious impact is that people are better connected to each other; for good and for bad. Some young people cannot put their phones down. They have become addicted to communications in many forms. Schools no longer teach kids cursive writing or how to add numbers in their heads. Schools no longer have books. That is just inconceivable to people of my generation. While we just used our heads, nowadays people get what they want using their gadgets.

What is also obvious is that the technology spawned by the space program has improved the lives of people all over the world. NASA promotes this with the Spinoffs program, and the effects are indeed global.

If there is an iconic image that represents the impact of the space program on society, one candidate is the image of astronauts on the Moon; another is the image of the Apollo Mission Control Center. People will be able to visit this iconic place and see for themselves where it all happened. It will be a grand experience, especially for youngsters. In years to come, one of those inspired kids may be the first human to set foot on Mars." – Manfred "Dutch" von Ehrenfried, Mercury, Gemini and Apollo Flight Controller.

As the control center is forever linked to the Apollo program itself, perhaps understanding the meaning of the Apollo Mission Control Center to the nation and the world is one and the same. Philosophers, be they poets, writers or seers, seem to say it best.

"To see the Earth as it truly is, small and blue and beautiful in that eternal silence where it floats, is to see ourselves as riders on the Earth together, brothers on that bright loveliness in the eternal cold; brothers who know now that they are truly brothers." – Archibald MacLeish, Librarian of Congress, Pulitzer Prize Winner, Poet.

Appendix 1
Key Correspondence

A number of letters were key to starting the restoration process.

Sandra J. Tetley, the JSC Historic Preservation Officer, wrote to Mark Wolfe, who was Executive Director State Historic Preservation Office, Texas Historical Commission (Figure A.1.1).

Eugene F. Kranz wrote to M. Wayne Donaldson, Chairman of the Advisory Council on Historic Preservation (Figure A.1.2).

The following article in the *Houston Chronicle* was instrumental in getting the Johnson Space Center management to agree to a meeting to discuss restoration of the control center.

Save Mission Control

The former nerve center for NASA put Houston on the map and should be restored

Copyright 2016: *Houston Chronicle*, December 20, 2016

One Sunday night in July a generation ago, the whole world was watching Houston.

On July 20, 1969, just before 10 p.m. local time, an awestruck television audience around the globe saw ghostly black-and-white images beamed back to a building at what was then called the Manned Spacecraft Center. Live from the lunar surface came the pictures and sounds of the first men setting foot on the moon. And families gathered around their television sets occasionally caught a wide-shot glimpse of Mission Control, the big room in Houston where hard-working people from the planet Earth made an impossible dream an improbable reality.

Today, just about anybody of a certain age can tell you where he was or what she was doing that night. Mission Control is arguably the most historically significant place in Houston, a designated National Historic Landmark. Yet this irreplaceable piece of our heritage has fallen into a sad state of disrepair.

© Springer International Publishing AG, part of Springer Nature 2018
M. von Ehrenfried, *Apollo Mission Control*, Springer Praxis Books,
https://doi.org/10.1007/978-3-319-76684-3

As the *Chronicle*'s Andrew Dansby reported ("Money, access complicate effort to restore Mission Control," Page A1, Dec. 4), the room that served as the nerve center for America's space triumphs during the Gemini, Apollo and early shuttle eras now sits badly neglected. Visitors have cut up the upholstery for souvenirs, ash tray covers in the viewing room have been pried loose and the carpeting is a mess. It's a dispiriting sight for former flight controllers, who decry the deterioration of this monument to our nation's space legacy.

Hordes of tourists taking daily Space Center Houston tours routinely visit the viewing area overlooking the room itself, ravaging the place where astronauts' families and other dignitaries watched through a window. Preservationists also complain way too many VIPs are allowed to walk amid the consoles inside Mission Control. The National Park Service reports that about 40,000 people a year are allowed to wander around this area. That number, the park service says, needs to drop to 2,000.

NASA and park service authorities have talked about restoring the room for decades, and space center officials have diligently preserved Apollo-era artifacts. A proposal released last year would cost an estimated $3 million and take about 18 months, but nobody's set a date for work to begin. A consultation meeting required by the National Historic Preservation Act has yet to happen.

And yes, somebody has to raise the money. Government employees aren't legally allowed to tackle this task, so private citizens need to step up to the plate. Space Center Houston, the nonprofit visitor center entity, recently hired a CEO with a fundraising background, but he'll need all the help he can get. Surely the city that landed a man on the moon can raise the money and get this restoration project off the ground. When the Saturn V rocket sitting on the periphery of Johnson Space Center fell into disrepair, a "Save the Saturn" drive raised $2.5 million by passing the hat around various government and private sources, preserving the moon rocket for future generations. Now Houston needs to save Mission Control.

The 50th anniversary of the Apollo 11 moon landing is now a little more than two years away. Less than a year later, NASA will mark the half-century anniversary of Apollo 13, the aborted moon mission depicted in the movie that made big-screen heroes out of the men working behind the scenes at the space center. So there's a sense of urgency about launching this restoration project as quickly as possible.

Our city has earned an unfortunate reputation for ignoring its past and allowing historic buildings to either fall into disrepair or fall to the wrecking ball. Mission Control, which did as much as anything to put Houston on the map in the 20th century, deserves better."

National Aeronautics and
Space Administration

Lyndon B. Johnson Space Center
2101 NASA Parkway
Houston, Texas 77058-3696

Reply to Attn of :
JP-17-022

Mr. Mark Wolfe
Executive Director
State Historic Preservation Office
Texas Historical Commission
P.O. Box 12276
Austin, TX 78711-2276

Subject: Section 106 Consultation – Determination of Effects
Restoration of the Apollo Mission Control Center and Viewing Room, a National
Historic Landmark, at the National Aeronautics and Space Administration (NASA)
Lyndon B. Johnson Space Center (JSC), Houston, Texas

Dear Mr. Wolfe:

In accordance with 36 CFR Part 800.10 (a), Protection of Historic Properties, Special
Requirements for protecting National Historic Landmarks; and Section 110 of the National
Historic Preservation Act of 1966, as amended through 2006 [54 U.S.C. 306107 "Planning
and actions to minimize harm to National Historic Landmarks (NHLs)"], NASA JSC would
like to present to the Texas Historical Commission our plans to restore the Apollo Mission
Control Center (MCC) including the Mission Operations Control Room (MOCR), Visitors
Viewing Room, Summary Display Projection Room (Bat Cave), Simulation Control Room
(SCR) and Recovery Operations Control Room (ROCR). Each of these rooms are in Building
30, Christopher C. Kraft, Jr. Mission Control Center, a National Historic Landmark.

On July 20, 2019, NASA and America will celebrate the 50th Anniversary of the first men to
land on the Moon. NASA JSC would like to celebrate that momentous and world course-
changing event with an accurately restored and engaging Apollo MCC from which visitors
will gain a sense of having experienced the feat that was the Apollo Program. One simply
needs to take a glimpse at the MCC to begin to experience a flood of emotion and American
pride, but upon closer inspection these feelings begin to fade as one observes the physical
condition of the historic furnishings that remain in the MOCR, Bat Cave, Viewing Room,
SCR, and ROCR. After years of visitors, unabridged access and declining budgets, the state
of the Apollo Mission Control Center National Historic Landmark declined to the point that
it was listed by the National Park Service (NPS) as "threatened" in September 2015. It is

Figure A.1.1 Letter from Sandra J. Tetley, JSC Historic Preservation Officer, to Mark Wolfe,
Executive Director State Historic Preservation Office, Texas Historical Commission.

JSC's intent to restore the Apollo MCC and develop a first class visitor experience to preserve and interpret the historic Apollo Program. (Enclosure 1)

In 2014, NASA JSC and the National Park Service (NPS) entered into an Interagency Agreement to prepare a Historic Furnishings Report and Visitor Experience Plan for the Apollo MCC NHL. This report and recommendations specifically focus on the MOCR, Bat Cave, Viewing Room, SCR, and ROCR. The report recommends restoration of the MCC, which will allow permanent retention of the existing Apollo Program materials and features while providing latitude to replace missing features and items. Once restored, these interior spaces can evoke powerful emotional and intellectual responses from visitors who can sense that "history happened here." The Historic Furnishings Report has served as the key planning tool in the proposed restoration of the MCC NHL.

The restoration of Apollo MCC will be completed following the recommendations in the Historic Furnishings Report and in accordance with the Secretary of the Interior's Standards for the Treatment of Historic Properties; therefore, NASA is committed to implementing this work consistent with 800.5(a)(2)(ii) to result in a "No Adverse Effect" to this historic property.

Furthermore, NASA intends to continue to seek review and comment from the SHPO and other consulting parties, to ensure the Secretary's Standards for the Treatment of Historic Properties are met as additional plans are finalized.

The following are the plans for the restoration of the Apollo MCC:

- Console Restoration - Restoration of all consoles to the Apollo era, specifically configure all consoles to mimic the Apollo 15 Operational Configuration, and ensure each console shell has an accurate array of hardwired devices appropriate for each position. The console arrays (panels, switches, indicators, monitors, etc.) will be configured to Apollo 15 locations, representing the apex of technological achievement of the Apollo Missions. (Enclosures 2 and 3) Console configurations will follow the "MCC Operation Configuration: Mission J1 Apollo 15" document (PHO-TR155, 03-26-71),
    - o For documentation purposes, interviews with Apollo Flight Controllers will be completed in the MOCR for each console before restoration activities begin to gain perspective and knowledge on activities at the console and personal items on console. These videos and oral histories will be preserved and placed on the JSC History website for public dissemination.
    - o Component restoration includes documentation of current condition, disassembling components, replication of components where needed, cleaning and preventative wax, reassembling components, and the curation of items not used. Facsimiles of components, component parts or console parts will be produced if necessary. All replicated components, component parts or console parts will be done using original or near original material. Facsimiles will accurately represent the original.
    - o Restoration will include cleaning corrosion and debris using mineral spirits and/or simple green. A Nilfisk vacuum, and Dremel or dental tools will be used as necessary.

(continued)

o This action also includes removal of console components and preparation for shipment to Cosmosphere where work will be performed; the identification, documentation, cataloguing, and labeling components and consoles; and packing (for curation purposes) any components not used in restoration.

o All components and modules not used in the restoration will be catalogued and preserved for storage.

o Complete documentation of the restoration process pre- and post-restoration will be included with this action item.

- Reanimation of consoles – Consoles and components will be reanimated at Cosmosphere. Working with the restoration team, NASA JSC, and NASA JSC Visitor Experience Team, the appropriate buttons, appropriate sequence, and monochromatic displays on the CRT monitors will be lighted. (Enclosures 2 and 3)

  o Lighting of components will consist of LED lighting systems behind components installed in a non-obtrusive, non-destructive manner and is removable.

  o Proper electrical wiring, junction boxes, and panels will be installed to support consoles, projectors, and other equipment in a non-obtrusive, non-destructive manner and is removable.

- Restoration of MOCR to approximate what it looked like at the moment the moon landing took place. (Enclosure 4)

  o For documentation purposes, 3-D Laser Scanning will be completed in the MOCR before and after restoration activities. The action item will include drawings to be included in a final report.

  o Acquisition and installation of appropriate quantity and array of personal items, such as ash trays, manuals, headsets, books, pencils, pencil sharpeners, clocks, tape dispensers, reel-to-reel tape players, maps, charts, coffee cups, documents and other objects among the console surfaces to recreate the historic scene during active Apollo Missions. Fast food boxes and other details added as well according to controller position.

    ▪ No authentic documents, maps, logs, manuals and objects to be placed on the consoles or in bookshelves will be used in this restoration. All original materials will be copied and/or replicated for this project.

  o Maps and other documents will be affixed to consoles so that they do not slide down and scratch consoles.

  o Acquisition and restoration or replication of missing furnishings such as ceiling-mounted group display cameras and tripod-mounted television cameras, waste receptacles, book cases, coat racks, period clothing and office supplies will be added to convey a period-specific, cohesive historic scene for visitors.

  o Steel bookcases will be filled with replicas of the 3-ring binders containing flight materials and other documents.

  o Installation of coat rack(s) with replicas of appropriate period clothing along south wall of MOCR.

  o Cleaning/repairing of existing gray vinyl chairs, which closely resemble those used during Apollo Missions. Only chairs that most accurately resemble those in historic photographs will be used.

(continued)

o   Recreation of wooden boxes that once sat behind consoles of various Flight
    Controllers as depicted in historic photographs and determined through
    personal interviews. (Enclosure 5)

o   Replacement of look-alike mission roses at the front of the MOCR under the
    display screens.

o   Removal and relocation of non-Apollo period objects including the podium,
    flag and microphone stand placed in the MOCR for ceremonies.

o   Reactivation of the large group displays on the west wall of the MOCR with
    appropriate projection technology to recreate Apollo-era use of the screens.
    Information displayed will be determined during restoration. Large group
    display screens will be cleaned and reconditioned.

o   Reactivation of group clocks (above front display screens) and a method to
    label and control them using LED lighting. Information displayed will be
    determined during restoration. Existing hardware will be reconditioned and
    new hardware installed as needed and in accordance with preservation
    requirements.

o   Restoration/repair of historic walnut railing on the north side of the consoles.

o   Removal of existing carpeting and replacement with replica gray carpet tiles
    that existed during Apollo Missions. Carpet selected is Mainstreet by
    Philadelphia, Capital III in the color Power Play, 80501.

o   Current lighting will be replaced with dimmable LED lamps, which will not
    produce UV to prevent further damage to the historic furnishings from long-
    term exposure to unfiltered UV radiation. Fixtures will remain intact, but
    ballasts will be retro-fitted to allow for installation of the LED lamps. Lighting
    will be diffused to ensure no hot spots and the lighting settings will create a
    similar interior environment as during the Apollo era which maximized
    control room operations. The historic "Blue" Overtone of lighting coming
    from World Map Screen will be recreated.

o   Restoration of Mission Plaques. This includes provision to inspect all of the
    mission plaques and provide restoration if needed. Shuttle Mission Plaques
    will be relocated into the north hallway of the MOCR.

o   Commemorative and memorial objects on the south MOCR wall will be
    retained in place. While these objects conflict with the restoration of the
    historic furnishings of the Apollo era, their symbolic significance means it is
    appropriate to keep these items in place as is.

o   Cleaning and repair of wall covering, HVAC grilles, and all surfaces. This
    action will also include vacuuming the ceiling and all surfaces.

o   Add non-obtrusive stanchions or barriers to the MOCR floor to prevent
    visitors from getting close to the consoles.

o   Recreation of the coffee bar that once sat in the north hallway of the MOCR
    including coffee maker, etc.

o   Install a see-through plexiglass panel to the opening of the lighting control
    area so Visitors to the MOCR floor can see the unique equipment.

o   Install a door from the bottom of the north steps exiting the Viewing Room
    into the MOCR for access to the MOCR floor. This will allow visitors to
    remain out of the Controlled Access Area of Building 30. Door will be
    secured with limited access to the floor. Additional details will be provided
    upon design completion.

(continued)

- Visitors Viewing Room – Restoration of the Visitors Viewing Room to approximate the period of significance which maximizes retention of the historic furnishings but also considers upkeep of the existing features that provide access based on ADA standards as they relate to historic properties. Retaining the elevator, ramp, and railings in the Visitors Viewing Area allows all members of the public to enter the NHL space through the historic primary entrance for visitors and view the MOCR from the historic vantage point. (Enclosures 6 and 7)
  - For the most part the proposed treatment for cleaning the wood, metal and fabric parts of each seat will be completed under the definition of "preservation". The materials and methods used to clean are not undertaken to make the seats "look new". The treatments are meant to only reduce the degradative particulates, soils, gum and grime found on the seats form many years of use. The reduction of these negative aspects will help in the long-term longevity and continued use of the seats. Only those aspects that affect the long term stability of the seats will be reduced and removed as necessary. The scars, kicks, gouges, etc. on these seats will be left in place as they show the history of the seats.

    The treatment proposed in regards to the functionality of the seats in general, is "restoration". Since the room and seats are utilized on a daily basis, restoration of the broken or loose seats will have to be undertaken. Parts that are broken or missing will have to be found or fabricated so that the seats will be usable. This is especially true in the case of the missing lids of the ashtrays. Seats that lean too far back or are unstable will have to be "restored" so that they can be safely used by the public.

  - Cleaning and repairing the existing theatre-style upholstered seating including seat mechanisms and ashtrays affixed to back of seats.
    - Cleaning of these historic pieces will be done using industry-accepted conservation materials and methods including brush vacuum, deionized water and/or a surfactant. Cleaning will be conducted to remove only surface staining and grime and will not restore the seats to their original condition. Research and examination of each seat will be conducted to determine the appropriate cleaning treatment and will be individualized to each seat.
    - Open seams in the fabric along the edges of the seat will be sewn closed suing black synthetic thread and conservation sewing techniques.
    - Seat mechanisms will be repaired/restored using found parts or replaced with current materials and methods if parts are not available.
    - Ashtrays on the backs of the seats will be repaired/restored using parts from like ashtrays or replaced with current materials and methods if parts are not available.
    - Evaluation of the wood end caps and back seats may determine that they are in good enough condition that a treatment of wet cleaning with a dilute detergent in aqueous solution will further reduce surface particulates, soiling, gum and grime. If so, this treatment will be conducted and then any residual solution will be removed using

(continued)

distilled water and a cotton cloth. A high quality conservation wax paste will then be applied to the wood portions of the seats and then lightly buffed with a cotton cloth.

o   Carpeting in the Visitors Viewing Area will be cleaned and restored.
    ▪   Damaged portion of carpet near phone booths will be restored using a portion of carpet harvested from under the desktop in one of the Communications booths.
    ▪   Carpet on wheelchair ramp will be replaced using carpet more in keeping with the period of significance.

o   Removal of the post-Apollo era photomurals from the walls and minimization of tour-related signage. All necessary tour-related signage will be displayed on removable tripods or other temporary display options. Repair wall covering as applicable.

o   Removal of window treatments in the Communications Booths in the Visitors Viewing Area.

o   Current lighting will be replaced with dimmable LED lamps, which will not produce UV to prevent further damage to the historic furnishings from long-term exposure to unfiltered UV radiation. Fixtures will remain intact, but ballasts will be retro-fitted to allow for installation of the LED lamps.

o   The torn and peeling edges of covering material on the standing desk in the back of the Visitors Viewing Room, which appears to be similar to or the same as the original wallcovering material in the viewing room and control room, will be repaired or recovered.
    ▪   Note: More research is required to determine if standing desk was originally covered or simply wood. If simply wood, wood will be refinished.

o   Black rotary telephone units will be added to the desks in the four telephone booths.

o   Security step lighting will be replaced with non-obtrusive lighting and electricity supply better hidden.

o   Non-Apollo era objects and furnishings will be removed from the Communications Booths. Photographs of the Communications Booths in use during a 1970 mission will be used to research the type and function of the desktop objects and source them for restoration of this space. Flight Controllers from the Apollo era will be consulted to determine the accuracy of the configuration drawing with respect to placement of the Audio-Visual Controller console.

o   Plywood shelf along west wall will be cleaned and all elements preserved; wires of audio jacks, electric, etc. will be secured and hidden; "reserved" plaques and jacks left in place.

o   Working replicas of 1960s TV will replace the current TVs which are not original.

o   Speakers that are hanging from the TV shelves which are not original will be removed.

o   Audio in Viewing Room will be restored or reactivated.

o   Interiors of telephone booths will be repainted in the original color.

(continued)

- The Summary Display Projection Room (the Bat Cave) will be restored to the period of significance as applicable to current use and to support Visitor Experience plans. (Enclosure 8 top)
  - o Establish which historic materials and furnishings are still in the room. An inventory of original equipment, materials and furnishings will be documented including all original equipment currently held in storage.
  - o Historic furnishings, objects and equipment that are in JSC storage shall be replaced in the Bat Cave for preservation purposes.
  - o Development of an interpretive display illustrating the function and capability of the Bat Cave will be developed.
  - o Projectors to project images and video on the applicable large displays will be installed.
  - o Audio system will be reactivated to support the Visitor Experience plans.

- The Simulation Control Room (SCR) will be restored to the period of significance. Restoration of this room will greatly improve the array of Apollo-era historic resources that are being preserved and interpreted for the public. Without at least one well-restored Staff Support Room, the representation of a functional matrix of physical spaces in Building 30 is lost.
  - o Remove curtains covering the glass window panes that allow visual access between the SCR and the MOCR. Repair wallpaper as necessary.
  - o Consoles that remain in the SCR will be arranged in a configuration that reflects the Apollo-era use of the room.
    - ▪ Modules will be restored and replaced as necessary to reflect the Apollo-era period of significance.
    - ▪ Consoles and modules will be treated in the same manner as the MOCR consoles.
  - o All non-historic objects from the SCR will be removed and stored elsewhere.
  - o The SCR will be cleaned thoroughly, including all objects that will be displayed there.
  - o Any items found in the SCR that cannot be used for the restoration will be retained and stored properly in the JSC storage facility.
  - o The lighting in the SCR will replicate historic operating levels during active use.
  - o The original "Selectomatic Transitubes" station in the SCR will be restored.
  - o Console chairs will be added to the consoles to complete the basic configuration of the room.
  - o Room will be painted in original color, damaged ceiling tiles replaced, and ceiling support grid painted.
  - o A "Sim Sup Escape Door" or replica thereof will be added in the appropriate location of the room. Additional details will be provided upon design completion.
  - o The door to the SCR will be changed to a glass door or a window installed into the existing door so Level 9 and VIP visitors can see into the room.

- The Recovery Operations Control Room (ROCR) is no longer available for restoration but a photo to visually mimic the ROCR will be placed in front of the original windows giving the appearance of the activity in the Apollo era ROCR.

(continued)

(Enclosure 8 bottom)

- Access to the MOCR floor has been controlled to prevent deterioration of the restored space and tours to the floor of the MOCR are now scheduled to avoid distraction to visitors in the Viewing Room. New Badge Readers on all doors have been installed with limited number of people given access.
  - o Visitors to the floor of the MOCR will be allowed limited access to area above the PAO console for photo opportunities.
  - o Visitors will not be allowed on the floor of the MOCR while there are visitors in the Viewing Room.
  - o Visitors will not be allowed near the consoles or to touch anything in the room.

Following the recommendations in the Historic Furnishings Report and as discussed in our Consultation Meeting in April, a Visitor Experience Plan is being developed for future visitor service operations, including the important stories to be told and how best to tell them. This plan is now being led by Space Center Houston and is currently under development. As details emerge, we will request comment on those plans.

During restoration, placards will be placed in the MOCR and Viewing Room describing the restoration in progress to Visitors. Our intent is to leave the MCC open to tourists and visitors and allow them to see the transformation of the NHL. Only when construction rules require an area to be closed to tourists will it be closed during construction.

A sustainment and maintenance plan is being developed with assistance from the National Park Service to assess the impact on the MCC from visitors to the historic site. A Visitors Effect Study will be developed to assess and monitor the impact of visitors to the MCC and will establish a periodic review of the effect to the NHL from visitation and develop a methodology for maintenance, repair and any changes to visitation. This monitoring will ensure the MCC is maintained, repaired as needed and sustained for the future. At the conclusion of the study, in consultation with the Consulting Parties, JSC will determine whether, and at what measurement interval, any additional future monitoring should occur.

JSC has determined that these restoration actions not only constitute "no adverse effect" but will remove the "Threatened" status designation of the NHL. Restoration will begin upon your concurrence and when funding is received. JSC intends to provide the Consulting Parties restoration status reports on a regular basis and we will request additional input when unforeseen issues and new considerations arise. Given the nature of this determination and our plans to restore this NHL, NASA JSC requests your concurrence to proceed.

Should you have any questions regarding this Consultation, please feel free to contact me at (281) 483-8113, or by email at sandra.j.tetley@nasa.gov. We look forward to working with you on this undertaking.

(continued)

JP-17-022   9

Sincerely,

Sandra J. Tetley
JSC Historic Preservation Officer

Enclosures:
1. Apollo Mission Control Center as it is today
2. Apollo MOCR during Apollo 11 and Apollo 15
3. Apollo MOCR during Apollo 15
4. Apollo MOCR after safe return of Apollo 11
5. Drawing of wooden boxes that sat behind Flight Controllers in MOCR
6. Visitors Viewing Room
7. Visitors Viewing Room
8. Summary Display Screens and Mission Clocks, Apollo 11

cc:
AD/S. White
AL/A. Xenofos
AL/T. Morrow
CD/T. LeBlanc
CD/J. Thornton
CD42/I. Moore
HQ/LD020/R. Klein
IS20/J. Ross-Nazzal
JA/S. Campbell
JA/J. Walker
JM/J. Haught
JM5/J. White
JM5/T. Pryor
JP/C. Noel
JP/R. Schlachter

Ms. Sharon Fleming, A. I. A, DSHPO
Director, Division of Architecture
Texas Historic Courthouse Preservation Program
Texas Historic Commission
P. O. Box 12276
Austin, Texas 76711-2276

(continued)

cc continued:

Ms. A. Elizabeth Brummett
State Coordinator for Project Review
Division of Architecture
Texas Historical Commission
P.O. Box 12276
Austin, Texas 78711-2276

Ms. Lydia Woods
Project Reviewer
Texas Historical Commission
P.O. Box 12276
Austin, TX 78711-2276

Mr. John Fowler
Executive Director
Advisory Council on Historic Preservation
401 F Street, NW
Suite 308
Washington, DC 20001-2637

Mr. Reid J. Nelson
Director
Office of Federal Agency Programs
Advisory Council on Historic Preservation
401 F Street, NW
Suite 308
Washington, DC 20001-2637

Mr. Tom McCulloch, Ph.D., R. P. A
Archaeologist
Advisory Council on Historic Preservation
401 F Street, NW
Suite 308
Washington, DC 20001-2637

Mr. Chris Daniel
Program Analyst
Advisory Council on Historic Preservation
401 F Street NW
Suite 308
Washington, DC 20001-2637

Ms. Amy Cole
Program Manager
Heritage Partnerships Program
National Park Service
Intermountain Regional Office
P.O. Box 25287
Denver, CO 80225-0287

Mr. Tom Keohan
Heritage Partnerships Program
National Park Service (NPS)
Cultural Resources IMDE-CNR
P.O. Box 25287
Denver, CO 80225-0287

Mr. Tracy Lamm
Chief Operating Officer
Space Center Houston
Manned Space Flight Education Foundation
1601 NASA Parkway
Houston, TX 77058

Mr. Eugene Kranz
Mr. Edward Fendell
Mr. Denny Holt
Mr. Jack Knight
Mr. Spencer Gardner

(continued)

January 26, 2017

Mr. M. Wayne Donaldson
Chairman
Advisory Council on Historic Preservation
401 F Street NW, Suite 308
Washington, DC. 20001-2637

Subject: Restoration of Historic Mission Control Center

Dear Mr. Donaldson:

I am asking your support to request the Acting NASA Administrator to save the Historic Mission Control Center.

For over 50 years, the NASA Johnson Space Center (JSC) in Houston has been considered the "home of human spaceflight" and, when Neil Armstrong landed on the Moon, he called the Mission Operations Control Room (MOCR) and his first words were, "Houston, the Eagle has landed."

I served in the NASA JSC Mission Control Center for over three decades. I began as a flight controller for Project Mercury, served as Flight Director during the Gemini and Apollo Programs, and as Director of Mission Operations for the early Space Shuttle missions (1960-1994). In my final years as the Director of Mission Operations, I was responsible for the MCC until my retirement in 1994. Two years later, after transfer of operations to a new Shuttle Operations Control Room, the original control room was abandoned. In 1985, the Apollo Mission Control Center was designated a National Historic Landmark.

For the past two decades, hundreds of thousands of visitors and have been through the viewing room, walked the floor of the historic MOCR and sat at the control consoles for pictures. The room has been used by employees to entertain friends, support social events, and as a short cut to other areas in the Mission Control Center. The Historic Mission Control Center consists of the following elements: MOCR, Viewing Room, Simulation Control Room (SCR), Recovery Operations Control Room (ROCR), and Bat Cave (large screen projection area).

The MOCR is in an embarrassing state, some clocks and console artifacts are missing, displays inert, carpeting torn and covered with duct tape, and the viewing room worn and dirty with pieces removed from desk coverings. The SCR, which is currently vacant, is in good shape and the ROCR is used for office space and storage. The overall condition is not emblematic of a National Historic Landmark.

Figure A.1.2 Letter from Eugene F. Kranz to M. Wayne Donaldson, Chairman of the Advisory Council on Historic Preservation.

In August 2014, I was advised that a report from The National Park Service (NPS) classified the Historic Mission Operations Control Center as a "threatened landmark," indicative of the poor stewardship by JSC management in the decades following its use after the Space Shuttle Program. I formed a small team of Apollo controllers and, for nearly three years, we have attempted to work with the JSC leadership to address its restoration. We have sent frequent communications to the current Center Director concurred upon by Dr. Christopher Kraft and Gerry Griffin, two former Center Directors, and by former Apollo Flight Directors. Our communications requested formation of a team of current and former stakeholders to lead the effort and collaboratively develop the plans to restore the MCC.

My team worked with the JSC Historic Preservation Officer and the NPS to produce the requirements for the Mission Control Center restoration consistent with policies for National Historic Landmarks. These requirements were used to develop cost, schedule, and an overall plan to complete the restoration by the 50th Anniversary of the Apollo 11 lunar landing, which will occur in two years.

In a July 2016 meeting, Space Center Houston (SCH), a 501c-3 entity, indicated that they would provide fund raising support for the restoration of the historic Mission Control Center. They were provided a copy of the restoration requirements we had developed with the NPS.

The JSC Director finally responded to our correspondence after a "Save Mission Control" editorial in the December 21, 2016, Houston Chronicle. The January 17, 2017, meeting with the JSC Center Director, SCH, and my Apollo team was not the working session I had expected. The meeting was not open for any discussion or any requirements from our side.

Mr. Jim Thornton, the JSC Restoration Project Manager, provided a two-page handout. The handout consisted of a Project Plan and Schedule. Restoration was limited to only the MOCR. There was no specification of the restoration requirements, standards, or indication of compliance with Federal standards. The Viewing Room, Bat Cave, and Lobby would be "updated." No details were provided on the proposed updates. The vacant SCR and the ROCR were excluded from restoration because of "operational constraints". There was no indication of an intent to use qualified restoration contractors for any of the proposed work. SCH was listed as providing funds for the above work.

The JSC meeting then focused on the projectors, speakers, lighting, and other elements related to Storytelling, which was also to be funded by SCH. My team had worked with SCH for over six months and this topic was never mentioned and it was never included in any of the discussions related to the fund raising. JSC indicated the Storytelling requirements were to be provided by SCH and JSC External Relations. I asked if the NPS would be included in developing the Storytelling. JSC indicated that this would be done in house.

Storytelling was a major JSC focus in previous years but was dropped because of cost estimates of up to $7 million. In addition, if JSC is proposing a similar system, it will significantly impact the integrity of the historic Mission Control Center. While no details were provided to us at the, meetings, the JSC Legal representative indicated knowledge of the proposed hardware.

(continued)

At the conclusion of the meeting, I was invited to participate in a weekly meeting and will have an Apollo controller attend as well.

Further detailed information on the JSC plans will be made available in three months. I attempted to obtain specific goals to be used in developing the detailed plans and none were provided. The meeting adjourned after about 35 minutes.

I and two others in this group have been deemed "Consulting Parties" by the Advisory Council on Historic Preservation (ACHP) per Section 106 of the National Historic Preservation Act (NHPA – P.L.108-72) and have the same standing with the NPS and the Texas State Historic Preservation Office. We want to ensure that this historical restoration is done, per NHPA standards, and that controllers will be available when needed to address specific configuration issues.

On July 20, 2019, our Nation will celebrate the golden anniversary of mankind's first voyage to and landing on Earth's natural satellite, the Moon. Against the background of three assassinations, an unpopular war in Southeast Asia, race riots, and antiwar protests, the lunar landing was the singular event that unified our Nation and galvanized the world. The Mission Control Center was critical and visible in the hours that followed while the world looked over our shoulders.

I believe that a fitting tribute to the early pioneers of space and those that will follow in their footsteps is the restoration of the historic Apollo Mission Control Center in time for the Golden Anniversary of the lunar landing, if not before.

Sincerely,

Eugene F. Kranz
Former NASA Flight Director
Consulting Party – NASA JSC Historic MCC Restoration Project

Cc: Robert Lightfoot, Acting NASA Administrator
       William Gerstenmaier, NASA Associate Administrator, Human Exploration & Ops.
       Dr. Ellen Ochoa, Director, NASA Johnson Space Center
       Mr. William T. Harris, President, Space Center Houston

Copies of this letter have been transmitted to The Texas State Historical Commission, National Park Service, Advisory Council on Historic Preservation, National Trust for Historic Preservation, U.S. Congressional Committees and Chairs, and NASA Federal Preservation Officer.

Appendix 2
Mission Manning Lists

These are the official manning lists from memoranda by the Director of Flight Operations to staff. There may be some individuals who are not listed because they were still in training or in positions that were not listed. In some cases, an individual who was "on job training" would not be listed.

Bill Reeves has taken on a labor of love with the intent of having a database of anyone who ever worked operations supporting the manned space flights that were flown out of the National Historic Landmark. He has taken all available historical information and created an Excel spreadsheet that goes further than the following lists in identifying individuals listed on official manning correspondence who served in the Apollo Mission Control Center in any capacity. This spreadsheet will be available on the MSOA website in the near future.

Note that the following lists only go up thru Skylab. The lists for the Apollo-Soyuz Test Project, Approach and Landing Tests and Skylab up thru STS-53 (the last mission flown out of MOCR 2 in 1992) are being compiled and be in the MSOA Manning Lists in the near future.

© Springer International Publishing AG, part of Springer Nature 2018
M. von Ehrenfried, *Apollo Mission Control*, Springer Praxis Books,
https://doi.org/10.1007/978-3-319-76684-3

Figure A.2.1   Flight Director Bill Reeves. Photo courtesy of NASA.

MERCURY

MERCURY CONTROL CENTER AND REMOTE SITE MANNING					"HAM"		SHEPARD	GRISSOM	
SITES/POSITION	MA-1 ######	MR-1 ######	MR-1A 12/19/1960	MR-2 1/31/1961	MA-2 2/21/1961	MA-3 4/24/1961	MR-3 5/5/1961	MR-4 7/21/1961	MA-4 9/13/1961
TEST NUMBER						#835	#108		
FLIGHT DIRECTOR	KRAFT	KRAFT	KRAFT	KRAFT	KRAFT	KRAFT	KRAFT KYLE	KRAFT	KRAFT
PROCEDURES		KRANZ JOHNSON	KRANZ	KRANZ JOHNSON	KRANZ	KRANZ	KRANZ HAVENSTEIN	KRANZ	KRANZ
SYSTEMS							ARABIAN		ARABIAN
CAPCOM							SLAYTON COOPER-B	SHEPARD	SHEPARD GLENN
							SLAYTON-B	GRISSOM
FLIGHT DYNAMICS		ROBERTS	ROBERTS	ROBERTS	ROBERTS	ROBERTS	ROBERTS	ROBERTS	ROBERTS
RETRO		HUSS	HUSS	HUSS	HUSS	HUSS	HUSS	HUSS	HUSS
ENVIRONMENT							SCHLER		SCHLER
SURGEON							WHITE		
NETWORK							CLEMENTS		
SCC							SHARKEY		
SIM SUP		FABER	FABER	FABER	FABER	FABER	FABER	FABER	FABER
BDA-FLT					BDA-FLT				HODGE
CC		SIM TASK GROUP 1959			CC				SLAYTON
FDO		FABER			FDO				LUNNEY
SYS		KOOS HOOVER			SYS				
ENV		LUNNEY MILLER			ENV				
SURGN					SURGN				
					CYI-CC	ERNULL		CYI-CC	BECKMAN
					SYS	BARKER		SYS	STENFORS
					A/M	WARD/BERRY		A/M	WARD
					KNO-CC	WAFFORD		KNO-CC	VOLPE
					SYS	N/A		SYS	HUBER
					A/M	N/A		A/M	KRATCHOVIL
					ZZB-CC	AIKENHEAD		ZZB-CC	LLEWELLYN
					SYS	N/A		SYS	WAFFORD
					A/M	N/A		A/M	FLOOD
					IOS-CC	N/A	LLEWELLYN		BRUMBERG
					SYS	N/A	RUMBAUGH		RUMBAUGH
					A/M	N/A	HALL		AUSTIN
					MUC-CC			MUC-CC	CARPENTER
					SYS			SYS	ROSENBLUTH
					A/M			A/M	BISHOP/TURNER
					WOM-CC	GUTHRIE		WOM-CC	GUTHRIE
					SYS	HOPP		SYS	BARKER
					A/M	LANE		A/M	LANE
					CTN-CC	VOLPE		CTN-CC	MOORE
					SYS	ROSENBLUTH		SYS	T. WHITE
					A/M			A/M	SMITH
					ATS-CC	MOORE	MOORE		LANGFORD
					STS	STENFORS	STENFORS		REMBERT
					A/M	AUSTIN/GORDON	AUSTIN/GORDON (Port CNV)		HAWKINS
					HAW-CC			HAW-CC	HIGGINS
					SYS			SYS	DELUCA
					A/M			A/M	MOSER
									COOPER?
					CAL-CC			CAL-CC	DURET
					SYS			SYS	LOONGAN
					A/M			A/M	PRUETT/MARCHBKS
						CARPENTER			SCHIRRA?
					GYM-CC	ALDRICH		GYM-CC	ALDRICH
					SYS	HUNTER		SYS	HUNTER
					A/M			A/M	DAVIS
					TEX-CC	BECKMAN		TEX-CC	FABER
					SYS	LONGAN		SYS	CROSS
					A/M	SMITH/HALL		A/M	HALL/HOLMSTROM

Figure A.2.2　Mercury Manning List.

MERCURY

SITE/POS'N	"ENOS" MA-5 11/29/1961	GLENN MA-6 2/20/1962	CARPENTER MA-7 5/24/1962	SCHIRRA MA-8 11/3/1962	COOPER MA-9 5/15/1963	
FLIGHT DIR	KRAFT	KRAFT	KRAFT	KRAFT	KRAFT	HODGE
PROCEDRS	KRANZ	KRANZ	KRANZ	KRANZ	VONEHREN	KRANZ
SYSTEMS	ARABIAN	ARABIAN	ARABIAN(?) ALDRICH	ALDRICH	ALDRICH LOCKARD	BROOKS GULL
CAPCOM	GRISSOM GLENN CARPENTER-B	SHEPARD SLAYTON CARPENTR-B	GRISSOM	SLAYTON	SCHIRRA	SHEPARD
FLIGHT DYN	ROBERTS	ROBERTS-LUNNEY	LUNNEY		LUNNEY	CHARLESWTH
RETRO	HUSS		LLEWELLYN-HUSS		LLEWELLYN	HUSS
ENVIRON	SCHLER		HUGHES-SAMONSKI		SAMONSKI	HUGHES
SURGEON			WHITE		BERRY	CATTERSON
NETWORK			BERRY		BURWELL	
SCC					HATCHER	SHARKEY
	FABER	HOOVER	MILLER/SULLIVAN	KOOS/BROOKS	BROOKS/FERGUSON	
BDA-FLT	HODGE	HODGE				
CC	SHEPARD	GRISSOM		TOMBERLIN		
FDO	LUNNEY	LUNNEY			BROWN	
SYS		TOMBERLIN STRICKLAND			MOSER	
ENV		SAMONSKI HUGHES		BROWN	SHEA	
SURG'N		BERRY COONS		KELLY		
CYI -CC	LANGFORD	LLEWELLYN	OLASKY	LEWIS	PLATT	
SYS	STENFORS	ROSENBLUTH	ROSENBLUTH	WAFFORD	REMBERT	
A/M	WARD/BURWELL	WARD/BURWELL	HAWKINS/UNGER	UNGER	KRATCHOVIL	
KNO- CC	LLEWELLYN	DURET	MOORE	HANNIGAN	PENDLEY	
SYS	ROSENBLUTH	RUMBAUGH	DELUCA	CASSELLI	BARKER	
A/M	KRATOCHVIL/MARCHBK	KRATOCHVIL/MARCHBKS	BECKMAN	BURWELL/BIDWELL	BLACKBURN	
ZZB - CC	BECKMAN	LANGFORD	LEWIS-PARKS	PARKS	TOMBERLIN	
SYS	LONGAN	REMBERT	WAFFORD	AMOS	ROSENBLUTH	H. SMITH
A/M	FLOOD/FOX	FLOOD/FOX	N/A		GRAVELINE	
CSQ-CC	HIGGINS	BECKMAN	DURET	LOCKARD	GLENN	
SYS	L. WHITE	HOPP	RUMBAUGH	T. WHITE	RUMBAUGH	
A/M	HALL/HANSEN	AUSTIN/BENSON	BENSON	KRATOCHVIL	BECKMAN	
		COOPER	SLAYTON			
MUC-CC	SCHIRRA-HOOVER	FABER	ERNULL	DURET	LEWIS	
SYS	BARKER	L. WHITE	HUNTER	HUNTER	STENFORS	
A/M	BECKMAN/BISHOP	BECKMAN/BISHOP	AUSTIN/BISHOP		BISHOP	
WOM - CC	VOLPE	VOLPE	BRUMBERG	FERGUSON		
SYS	WAFFORD	WAFFORD	HUBER	DUNBAR		
A/M	LANE/OVERHOLT	OVERHOLT/LANE	LANE	LANE		
CTN-CC	OLASKY	HIGGINS	GUTHRIE	PLATT	PARKS	
SYS	DELUCA	HUBER	CROSS	RUMBAUGH	AMOS	
A/M	GRAVELINE/HOLMSTROM	GRAVELINE/HOLMSTROM	FLOOD	SHEA	LUCHINA	
RKV-CC	ZEDEKAR	PRIM		PCS-SHEPARD	HUNTER	
SYS	HUBER	CROSS		STRICKLAND	WALSH	
A/M	KELLY/HAWKINS	R. KELLY/ F/ KELLY		BECKMAN	BRATT	
HAW-CC	ERNULL	ERNULL-OLASKY	PRIM	GRISSOM	CARPENTER	ROACH
SYS	T. WHITE	LONGAN	REMBERT	BARKER	STRICKLAND	DAVIS
A/M	MOSER/AUSTIN	MOSER/HALL	MOSER	MOSER	MOSER	
	COOPER	SCHIRRA		GLENN		
CAL-CC	ALDRICH/PRIM	ALDRICH	SHEPARD	PRIM	T. WHITE	
SYS	REMBERT	T. WHITE	T. WHITE	REMBERT	DUNBAR	
A/M	BRATT/PRUETT/BENSON	BRATT/PRUETT	GRAVELINE/REED	HALL/TUCHINA	KELLY	
	SLAYTON			CARPENTER		
GYM-CC	MOORE	MOORE-BRUMBERG	COOPER	KAWALKIEWICZ	GRISSOM	
SYS	HUNTER	HUNTER	BARKER	ROSENBLUTH	WAFFORD	
A/M	DAVIS/TURNER	DAVIS/TURNER	REED/HOLMSTROM	GRASVELINE/MUSGRAVE	G. SMITH	
TEX-CC	KUEHNEL		LANGFORD/WHITE	MUSE-BATES-ROACH	BRAY	
SYS	RUMBAUGH		MUSE/BROWN	L. WHITE	L. WHITE	
A/M	SMITH/KELLY/LAWSON		PLATT/GULL	BLACKBURN	BENSON	
			KAWALWIEWICZ			
				WATERTOWN-GULL		

GEMINI FLIGHT CONTROLLER MANNING

SITE/POSITION	GT-1	NS-1	GEMINI 2	GEMINI 3	GEMINI 4	SEE CAPE TEAM AS BACKUP	
	4/8/1964	10/9-17/64	1/19/1965	3/23/1965	6/3/65	MCDIVITT-WHITE	
FLIGHT		KRAFT	KRAFT	KRAFT	KRAFT	KRANZ	HODGE
AFD		KRANZ	KRANZ	KRANZ	PLATT	VONEHRENFRIED	HARLAN
O&P		ROACH	VONEHRENFRIED	VONEHRENFRIED	ROACH	TOMBERLIN	ARMSTRONG
			ROACH	ARMSTRONG			
SURGEON		BERRY	BERRY	BERRY	BERRY	CATTERSON	COONS
		COONS	COONS	COONS			
CAPCOM		COOPER	COOPER	COOPER	GRISSOM	CHAFFEE	CERNAN
				CHAFFEE(HUS)	WILLIAM(CAPE)		
MMC MONITOR		GOODKIND	GOODKIND	GOODKIND	GOODKIND		
BOOSTER		PLATT	PLATT	PLATT	PLATT		
TANKS		FREEMAN	CERNAN	CERNAN	CERNAN		
SWITCHOVER		PARKER-CARTER	PARKER-CARTER	PARKER-CARTER			
GNC		ALDRICH	ALDRICH	ALDRICH	ALDRICH	COEN	GRIFFIN
		LOCKARD	GRIFFIN	GRIFFIN			
EECOM		T. WHITE	GLOVER	GLOVER	AARON	GLOVER	LOE
		GLOVER	AARON-BELL	AARON-BELL	BELL		
SYSTEM SUPPORT			BOATMAN	BOATMAN			
			TUNNICLIFF	TUNNICLIFF			
AGENA							
AGENA SYS							
FDO		LUNNEY	CHARLESWORTH	CHARLESWORTH	CHARLESWORTH	PAVELKA	BOSTICK
		CHARLESWORTH	LUNNEY	LUNNEY			
RETRO		LLEWELLYN	LLEWELLYN	LLEWELLYN	LLEWELLYN	CARTER	MASSARO
		BOSTICK	BOSTICK	BOSTICK			
NETWORK		SHERIDAN	SHERIDAN		PISKE	ARELLANO	NICKERSON
M&O					HATCHER	JONES	EGAN
SIMSUP			FERGUSON	FERGUSON	MILLER/FERGUSON		
SIM COORD					BATES		
PAO							
CYI - CC		ROY-DAVIS		ROY-DAVIS	EALICK		
GEMINI		McGHEE		COEN-BARKER	MOSER		
AGENA				SAULTZ A/S	PRESENT A/S		
A/M		POLLARD		BECKMAN	SHAMBUREK		
CRO - CC		HUNTER-CONRAD		HUNTER-CONRAD	FENDELL		
GEMINI		MUSE-LOE-WALTON		WHITE	SHITH-FULLER		
AGENA		SMITH		WALTON A/S	FERRY A/S		
A/M				POLLARD-BISHOP	POLARD+2AUS		
OBSERVERS					D. SCOTT		
HAW -CC		TOMBERLIN-EALICK		TOMBERLIN-EALIC	A DAVIS		
GEMINI		MOSER-HORTON		MOSER-LINK	BARKER		
AGENA				FUCCI A/S	W. YOUNG A/S		
A/M		YOUNG		CATTERSON-UNG	JERNIGAN		
OBSERVERS					CUNNINGHAM		
GYM - CC		KUNDEL-WAFFORD		LEWIS-SCOTT	GARVIN		
GEMINI		KLINGBEIL-PETERS		MUSE-STEPHENS	WALSH		
AGENA				DUNBAR A/S	M LOWE A/S		
A/M		CATTERSON-UNGER		BURWELL	WALMSLEY		
				ARMSTRONG	ANDERS		
CSQ - CC		GARVIN-LEWIS	LEWIS-GARVIN	GARVIN-ROSENBL	LEWIS-DRAUGHON		
GEMINI		WALSH-MIDDLETON	WALSH-FULLER	WALSH-CLAUNCH	T.WHITE-LINK		
AGENA				PERKINS A/S	BORCHES A/S		
A/M		PRUETT-JERNIGAN		JERNIGAN	HUMBERT		
RKV - CC		FENDELL-SCOTT	FENDELL-SCOTT	FENDELL	KUNDEL		
GEMINI		COEN-BARKER	BARKER-CLAUNCH	SMITH-FULLER	MUSE-BLISS		
AGENA			PAYNE(SIM)	BERLIN A/S	RUSSEL A/S		
A/M		KELLY-GRAVELINE		KELLY-GRAVEL;IN	KELLY		
TEX - CC		CHANDLER		KUNDEL	SCOTT		
GEMINI		LINK-AARON		KLINGBEIL-BLISS	McGHEE		
AGENA				EMERSON A/S	KEYSER A/S		
A/M		HUMBERT-HAWKINS		CHUBB	GRAVELINE		

Figure A.2.3 Gemini Manning List.

SITE/POSITION	GEMINI 5 8/21-29/65 COOPER-CONRAD			GEMINI 6 SC REMOTE SITE MANNING	GEMINI 76 12/15/665 SCHIRRA-STAFFORD GEMINI BORMAN-LOVELL 12/4-18/65		
FLIGHT	KRAFT	KRANZ	HODGE		KRAFT	KRANZ	HODGE
AFD	PLATT	VONEHRENFR	HARLAN		HARLAN	VONEHRENFRD	PLATT
						SCOTT	MERCIER
O&P	ROACH	ARMSTRONG	MOLNAR		ROACH	ARMSTRONG	MOLNAR
SURGEON	BERRY	CATTERSON	KELLY		BERRY	CATTERSON	COONS
CAPCOM	McDIVITT GRISSOM(CAPE)	ALDRIN ARMSTRONG	SCOTT		SEE-YOUNG BEAN-(CAPE)	CERNAN	BASSETT GRISSOM
MMC MON							
BOOSTER					HARLAN		
TANKS	BASSETT				WILLIAM		
GUIDANCE	PARKER	FENNER	RUSSELL		PARKER	FENNER	RUSSELL
GNC	GRIFFIN	STEPHENSON	ALDRICH		ALDRICH	GRIFFIN	COEN
EECOM	GLOVER	AARON	MERRITT		GLOVER	LOE	AARON
SPAN	BOATMAN	TUNNICLIFF	BROWN				
AGENA/NAV					BROOKS	WALTON	
AGENA SYS					SAULTZ	CARLTON	
FDO	BOSTICK	CHARLESWORTH	PAVELKA		CHARLESWORTH	PAVELKA	BOSTICK
RETRO	CARTER	MASSARO	LLEWELLYN		LLEWELLYN	MASSARO	CARTER
NETWORK	PISKE	NICKERSON	ARRELLANO		ARELLANO	RANDALL	P[ISKE
M&O	HATCHER	JONES	EGAN		HATCHER	CARR	DYE
SIMSUP	SHELLEY PAYNE			FERGUSON	SHELLEY/GRIFFITH HOLKAN		
PAO	HANEY	CHOP	WHITE		HANEY	CHOP	WHITE
CYI - CC	KUNDEL			ROY	FUCCI		
GEMINI	CLAUNCH			FULLER	G6-CLAUNCH		
AGENA	BURTON A/S			GRUBY-CONTOIS	G-7 LEGLER, ESPINOZA		
A/M	NUGENT			SHAMBUREK	ORD, WILSON		
CRO - CC	LEWIS			GARVIN-DRAUGHON	KUNDEL		
GEMINI	KLINGBEIL			T. WHITE	G6 MOSER		
AGENA	CANIN A/S			SMITH-WEICHEL	G7BLISS A/S DUNBAR		
A/M	JERNIGAN			HUMBERT-BISHOP	BECKMAN, MURY, WALCH		
OBSERVERS							
HAW -CC	GARVIN			LEWIS-DAVIS	FENDELL, BUCHHOLZ		
GEMINI	WALSH			KLINGBEIL	G6- PERKINS		
AGENA	EDELIN A/S			PERKINS-NERING	G7- FULLER A/S COLLINS		
A/M	ENDERS			SAWYER	PRESCOTT, WALMSLEY		
OBSERVERS							
GYM - CC	FENDELL			KUNDEL-BUCHOLZ	SCOTT		
GEMINI	FULLER			MUSE-LINK	G6- CONWAY		
AGENA	WATROS A/S			ROBINSON	G7- BARKER A/S BERL;IN		
A/M	ORD			WALTER	CHUBB, NELSON		
CSQ - CC	ROY			FENDELL	LEWIS, DRAUGHON		
GEMINI	T. WHITE-CONWAY			MOSER	G6- MUSE		
AGENA	PERRY A/S			BERLIN-EMERSON	G7-SMITH A/S ROBINSON		
A/M	CHUBB-EWING			WALMSLEY	IONNO, MAHER		
RKV - CC	SCOTT-FUCCI			FUCCI	GARVIN		
GEMINI	BARKER-BLISS			CLAUNCH	G6- LINK		
AGENA	HARMAN A/S			LEGLER-ESPINOZA	G7- WALSH A/S J. BATES		
A/M	NORSS			YOUNG-GOSSETT	BARRETT, CARVER		
TEX - CC	TRAINING SITE			TRAINING SITE	ROY, BASTEDO		
GEMINI				BARKER	KLINGBEIL		
AGENA							
A/M					ENDERS		

(continued)

SITE/POSITION	GEMINI 8		GEMINI 9			GEMINI 10		
	3/16/66 ARMSTRONG-SCOTT		6/3-6/66 STAFFORD-CERNAN		GEMINI	7/18-21/66 YOUNG-COLLINS		
FLIGHT	HODGE-CEC	KRANZ	KRANZ	LUNNEY	CHARLESWORTH	LUNNEY		CHARLESWORTH
AFD	PLATT	BUTLER	ROACH	BUTLER	PLATT	ROACH	HARLAN	BUTLER
	PRESENT	LEWIS	FENDELL		CANIN			
O&P	ARMSTRONG	BRITTON	MOLNAR	BATES	SUTTON	BRITTON	KEYSER	D. HOLKAN
	D. HOLKAN	LOWE	FISHER					
SURGEON			BERRY	COONS	CATTERSON KELLY	BERRY	CATTERSON	KELLY
CAPCOM	LOVELL-ANDER BEAN		ARMSTRONG	GORDON		COOPER		ALDRIN
	CUNNING(CAPE)		LOVELL	ALDRIN	ALDRIN(CAPE)	COOPER(CAPE)		
MMC MON								
BOOSTER	HARLAN		PLATT			BUTLER		
TANKS	SCHWEIKART		COOPER			CARPENTER		
GUIDANCE	PARKER	FENNER	FENNER	RUSSELL	PARKER	FENNER	RUSSELL	BALES
			BALES					
GNC	COEN	GRIFFIN	GRIFFIN	ALDRICH	COEN	COEN	GRIFFIN WILLOUGHBY	ALDRICH
EECOM	GLOVER	LOE	AARON	MERRITT	LOE	AARON	GLOVER	HOWARD
						BURTON	DELMONT	T. LOWE
SPAN			BOATMAN	TUNNICLIFF	BROWN	BOATMAN	TUNNICLIFF	BROWN
AGENA/NAV	BROOKS	SAULTZ	PUDDY(TGT)	HANNIGAN(TGT)		BROOKS	WALTON	SAULTZ
AGENA SYS	WALTON	CARLTON	SAULTZ(TGT)	WALTON(TGT)		LODEN	BRABANT	CARLTON
FDO	BOSTICK	PAVELKA	PAVELKA	DAVIS	BOSTICK	BOSTICK	DAVIS	PAVELKA
RETRO	CARTER	MASSARO	MASSARO	GRAVETT LLEWELLYN	CARTER	CARTER	GRAVETT	MASSARO
NETWORK	RANDALL	LONERO	PISKE	WHITE	JENKINS	RANDALL		OJALEHTO
M&O	HATCHER	EGAN	WILSON	EGAN	McLAUGHLIN	WHITE	PITTMAN	METCALF
SIMSUP	GRIFFITH/SHELLEY		GRIFFITH/SHELLEY			GRIFFITH/SHELLEY		
PAO	HANEY	CHOP	HANEY	RIEM/BOND	WHITE	HANEY	REIM	WHITE
			CHOP	RILEY	WEIL	CHOP-WORRELL	BOND-RILEY	WEIL
CYI - CC	GARVIN-BASTEDO		DRAUGHON			BASTEDO		
GEMINI	T. WHITE		WALSH			T. WHITE		
AGENA	BERLIN-ESPINOZA		ROBINSON/KNIGHT			LEGLER-LANGENFIELD		
A/M	SAWYER		WILSON			???		
CRO - CC	FENDELL		GARVIN-BRIZZOLARA			FUCCI		
GEMINI	CLAUNCH		BARKER-DIGENOVA			WALSH		
AGENA	SMITH-WEICHEL		PERKINS/PUDDY			NERING-SMITH		
A/M	BISHOP		WALSH			MORRISON		
OBSERVERS								
HAW -CC	ROY-MOLNAR		SCOTT-ARMSTRONG			FENDELL		
GEMINI	MUSE-CONWAY		CLAUNCH/FRANKLIN			BARKER		
AGENA	LEGLER-ROBINSON		CONTOIS/WEICHEL			CONTOIS-STELL		
A/M	DROESCHER		CONLEY/NUGENT			BARRETT		
OBSERVERS								
GYM - CC	DRAUGHON-BRIZZOLARA					DRAUGHON		
GEMINI	BLISS					MOSER		
AGENA	CONTOIS					EXPINOZA-JOKI		
A/M	GOSSETT					MILLINGTON		
CSQ - CC	FUCCI-BUCHOLZ		BUCHHOLZ			SCOTT		
GEMINI	BARKER		FULLER-BURRILL			CLAUNCH		
AGENA	PERKINS-NERING		LEGLER			BERLIN-FLATT		
A/M	WALTER		ENDERS			YOUNG		
RKV - CC	KUNDEL		KUNDEL-BASTEDO			BUCHHOLZ		
GEMINI	WALSH		MUSE-POLMANTEER			CONWAY		
AGENA	GRUBY-MOON		ESPINOZA			ROBINSON		
A/M	NUGENT		WITTMER			STEVENSON		
TEX - CC	TRAINING SITE		GRUBY			LINK -CNV		
GEMINI								
AGENA								
A/M								

(continued)

SITE/POSITION	GEMINI 11			GEMINI 12		
	9/12-15/66	CONRAD-GORDON		11/11-15/66 GEMINI	LOVELL-ALDRIN	
FLIGHT	CHARLESWORTH		LUNNEY	LUNNEY	CHARLESWORTH	KRANZ
AFD	BATES	PLATT	BUTLER	BUTLER	PLATT	
O&P	SUTTON	D. HOLKAN	FISHER	SUTTON	BRITTON	HOLKAN
SURGEON	COONS	ZIEGELSHCMID	CATTERSON	COONS	KELLY	HAWKINS
				CARPENTIER	DROESCHER	
CAPCOM	YOUNG		BEAN	CONRAD	ANDERS	
	WILLIAMSD(CAPE)			ROSSA (CAPE)		
MMC MON						
BOOSTER	BUTLER			PLATT		
TANKS	CARPENTER			DUKE		
GUIDANCE	BALES	FENNER	RUSSELL	FENNER-LONG	BALES	RUSSELL
GNC	COEN	ALDRICH	GRIFFIN	COEN	BATSON	
			BATSON			
EECOM	AARON	GLOVER	LOE	AARON	MERRITT	
	HOWARD	DELMONT	BURTON			
SPAN				LOE	GRIFFIN	ALDRICH
				BATES-EXP	FISHER-EXP	
AGENA/NAV	WALTON	SAULTZ	BROOKS	WALTON	BROOKS-SAULTZ	PETERS
AGENA SYS	BRABANDT	CARLTON	LODEN	BRABANT	CARLTON	LODEN
FDO	PAVELKA	BOSTICK	DAVIS	PAVELKA	DAVIS	BOSTICK
	KENNEDY					
RETRO	MASSARO	GRAVETT	CARTER	GRAVETT-CARTER	MASSARO	LLEWELLYN
				HOLLOWAY-FAO	NETTLES-FAO	
NETWORK	WHITE	MONKVIC	OJALEHTO	WHITE	OJALEHTO	MONKVIC
M&O	EGAN	WILSON	COLLINS	PHILLIPS	WILSON	JONES
SIMSUP	GRIFFITH			GRIFFITH		
PAO	HANEY	MCLEISH	WORRELL	WHITE-BOND	MCLEISH	JAMES
	BOND-REIM	WEIL	RILEY	REIM	WEIL	WORRELL
CYI - CC	BUCHHOLZ			BASTEDO		
GEMINI	BARKER			CONWAY		
AGENA	WEICHEL			NERING		
A/M	CARPENTIER			MILLINGTON		
CRO - CC	GARVIN			FUCCI		BERLIN
GEMINI	CONWAY			WALSH		
AGENA	GRUBY			BERLIN		
A/M	REED			WALSH-REED		
OBSERVERS						
HAW -CC	FUCCI			KUNDEL		
GEMINI	CLAUNCH			BARKER		
AGENA	PERKINS			WEICHEL		
A/M	WITTMER			STEPHENSON		
OBSERVERS						
GYM - CC						
GEMINI						
AGENA						
A/M						
CSQ - CC	BASTEDO			GARVIN		
GEMINI	T. WHITE			BLISS		
AGENA	LEGLER			PERKINS		
A/M	JONES			SAWYER		
RKV - CC	KUNDEL			BUCHOLZ		
GEMINI	WALSH			LEGLER		
AGENA	BERLIN			GRUBY		
A/M	NUGENT			JONES		
TEX - CC	LINK-CNV					
GEMINI						
AGENA						
A/M						

APOLLO

APOLLO FLIGHT CONTROLLER MANNING						
POSITION	AS201	AS202	AS203	NS-2		
	2/26/66 1112AM	8/25/66 1:15PM	7/5/66 10:53AM	Oct-66		
	SIB CSM-009	SIB S/C011	SIB/NOSE CONE			
	LC34	LC34	LC37B			
MISSION DIR				BOLENDER	HOLCOMB	ALLER
FOD						
FLIGHT DIR	LUNNEY	HODGE	HODGE	KRAFT	HODGE	KRANZ
AFD	CANIN	PRESENT	PRESENT	HARLAN	ROACH	PRESENT
CAPCOMS						
CSM/GNC	DELUCA	GRIFFIN	N/A	ALDRICH	GRIFFIN	WILLOUGHBY
		DELUCA				
CSM EECOM	PENDLEY	GLOVER	N/A	GLOVER	BURTON	LOE
		BURTON				
INCO						
FIDO	G. MEYER	G. MEYER	G. MEYER	SHAFFER	REED	KENNEDY
			KENNEDY	BOSTICK		
RETRO	DEITERICH	LLEWELLYN	N/A			
		DEITERICH	PAULES			
GUIDO	GUTHRIE	PARKER		PRESLEY	VONEHRENFRIED	WELLS
	PAULES	PAULES		PAULES		
		RENICK				
BSE-1	CASEY	CASEY	CASEY	CASEY		
BSE-2	BURDSHAW	BURDSHAW	BURDSHAW	BURDESHAW		
BSE-3		WOLF	WOLF	WOLF		
LM CONTROL						
LM TELCOM						
LM EMU						
SURGEON			N/A	CATTERSON	KELLY	COONS
				CARPENTIER	BECKMAN	
SIMSUP	KOOS	KOOS				
SIM COORD	HUNTER(AF)	CASEY(AF)				
EXPERIMENTS			N/A			
NETWORK	PISKE	LONERO	RANDALL	RANDALL	LONERO	AYERS
SCC	CHILDRESS	BARNES	CHILDRESS	HATCHER	WILSON	EGAN
M&O	CAGEL	JONES	CAGLE			
PROCEDURES	TEMPLE	ARMSTRONG	LOWE	TEMPLE	KEYSER	MOLNAR
	PORTER	BRIZZOLARO	GOODWIN			
		LIEBERGOTT	MERCIER			
FLT ACT OFF				ANDERSON	COTTER	JONES
PAO	HANEY	HANEY	CHOP	HANEY	JAMES	RILEY
PAO ASST		BOND				
PHOTO		MONCRIEF	PATNESKY			
ARTIST						
SIMULATION	KOOS	KOOS				
SPAN		O. MAYNARD				
RKV-VAN CC	EALICK	ROY		FENDELL		
	PRESENT			DRAUGHON		
EECOM	BLAIR	MOON		BURRILL		
GNC	WILLOUGHBY	BORCHESS		MUSE		
BSE	HAYES	MCCUMBER		MCCUMBER		
LM				H. SMITH		
SURGN				BARRETT-ADEED-ZIEGLESCHMID		
CRO- CC		EALICK	ROY	LEWIS-EALICK		
EECOM		SULMEISTERS		GARDNER-POLMANTEER		
GNC		FERRY		FULLER-FRANKLIN		
BSE/FIDO		GUTHRIE	HAYES	DAY		
LM				CONTOIS		
SURGEON				BEERS-AUSDTIN-WALSH		
CSQ -CC		CANIN				
EECOM		GARDNER				
GNC		KALISHEK				
GYM -CC			CANIN	DAVIS-ROY		
EECOM				WAFFORD		
GNC				PRINGLE		
BSE			DAY	HAYES		
LM				ESPINOZA		
SURGN				DROESCHER-FERGUSON-SAWYER		
BDA- CC			EALICK			
CYI-CC				SCOTT-CANIN		
EECOM				MOON-CLAUNCH		
GNC				DIGENOVA-Gf ASMEDER		
SURGEON				WILSON-HODGSON-CHAPPELL		

Figure A.2.4 Apollo Manning List.

APOLLO

POSITION	APOLLO 1/AS204			APOLLO5/AS204L		APOLLO 4/AS501	
	1/27/1967			1/22/68 5:48P LC37B		11/9/67 7:00AM	
	GRISSOM, WHITE, CHAFFEE			SIB LM-1/NOSE CONE		LC 39A	
MISSION DIR	BOLENDER	LEE	HOLCOMB	SCHNEIDER	LEE	SCXHNEIDEI	LEE
FOD				KRAFT		KRAFT	
FLIGHT DIR	KRAFT	KRANZ	HODGE	KRANZ	HODGE	CLIFF CHAS	LUNNEY
				LUNNEY B/U		PLATT	LIEBERGOTT
AFD	HARLAN	CANIN	ROACH	EALICK	LEWIS		
CAPCOMS	LOVELL	ARMSTRONG	BEAN				
CSM/GNC	GRIFFIN	WILLOUGHBY	ALDRICH			GRIFFIN	WILOUGHBY
							BORCHESS
CSM EECOM	GLOVER	LOE	BURTON			HOWARD	BURTON
			AARON				
INCO							
FIDO	SHAFFER	KENNEDY	REED	REED	KENNEDY		
	BOSTICK			BOSTICK	STOVAL		
RETRO	LLEWELLYN	DEITERICH	PAYNE	A'ANSON	SPENCER	LLEWELLYN	PAYNE
				LLEWELLYN		GRAVETT	
GUIDO	PRESLEY	WELLS	VONEHRENFRIED	RENICK	RUSSELL	HUTCHINSO	PAULES
	PAULES			PARKER	FENNER		BALES
BSE-1	CASEY			BRADY		WOLF-CASEY	
BSE-2	BURDSHAW			HOOPER		AMOS	
BSE-3	WOLF			CASEY-B/U			
LM CONTROL				CRAVEN	CARLTON		
LM TELCOM				PUDDY	MERRITT		
LM EMU							
SURGEON	BERRY	COONS	CATTERSON				
SIMSUP				GRIFFITH		KOOS	
SIM COORD				SMITH(AF)		CASEY(AF)	
EXPERIMENTS	BATES	FISHER	LOWE				
NETWORK	RANDALL	AYERS	LONERO	OJALEHTO	PHILLIPS	LONERO	WILSON
SCC	HATCHER	EGAN	WILSON	GUY	WILSON		
M&O							
PROCEDURES	MOLNAR	KEYSER	TEMPLE	GOODWIN	ARMSTRONG	HERVEY	HOLKAN
FLT ACT OFF	ANDERSON	JONES	COTTER				
PAO	HANEY	JAMES	WHITE	GREEN	WHITE	JAMES	HANEY
PAO ASST	BOND	WEIL	WORRELL	BOND	WORRELL	BOND	WEIL
PHOTO							
ARTIST							
SIMULATION							
SPAN				SAULTZ	HANNIGAN		
RKV CC	SCOTT			CANIN			
EECOM	MOON						
GNC	DIGENOVA						
BSE							
LM				GRUBY-THORSON			
SURGN	BARRETT-DAVIS						
CRO- CC	LEWIS			SCOTT		FENDELL	
EECOM	POLMANTEER					DEATKINE	
GNC	FULLER					KALISHEK	
BSE/FIDO	DAY			PARK-HOFFMAN		DAY-HOFFMAN	
LM				LODEN-PERKINS		CTL-STRAHLE	
SURGEON	WALSH-BROWN						
CSQ -CC	DRAUGHON			FUCCI			
EECOM	GARDNER						
GNC	STRAHLE			LM-LEGLER-HESELMEYER		REMOTED SITES MANNING	
SURGEON	WHITMER-KAPLAN			BSE-HAYES-HOFFMAN		GWM-HARLAN	
GYM -CC	A.DAVIS					ASN-DRAUGHON	
EECOM	BURRILL-CLAUNCH					HAW-MOLNAR	
GNC	MUSE					GYM-SUTTON	
BSE	HAYES					ARIA CAPE-KUNDEL	
LM							
SURGN	HENDERSON-ROWDEN						
BDA- CC							
VAN-CC	FENDELL					BASTEDO	
EECOM	WAFFORD-DEATKINE					CONWAY	
GNC	GRASMEDER					DIGENOVA	
SURGEON	HUMBERT-YOUNGMAN					CTL-KAMMAN	
						FDO-GREENE	
						BSE-MCCUMBER-PUSCHINSKY	

(continued)

APOLLO

POSITION	APOLLO 6/AS502		APOLLO 7			APOLLO 8		
	4/4/68 7:00:01 EST		10/11/68 1102 AM			12/21/68 7:51AM		
	CM-020,SM-014, LTA-2R		SCHIRRA-CUNNINGHAM-EISELE			BORMAN-LOVELL-ANDERS		
			LC34D CSM 101			LC39A CSM103		
MISSION DIR	SCHNEIDER	LEE	SCHNEIDER	LEE	MCMULLEN	SCHNEIDER	LEE	MCMULLEN
FOD	KRAFT		KRAFT			KRAFT		
FLIGHT DIR	CLIFF C		LUNNEY	KRANZ	GRIFFIN	CLIFF C	LUNNEY	WINDLER
AFD	PLATT	ROACH	FENDELL	EALICK	HARLAN	DRAUGHON	PLATT	EALICK
						ROACH		HARLAN
CAPCOMS			STAFFORD	EVANS	POGUE	COLLINS	MATTINGLY	CARR
			SWIGERT	YOUNG	CERNAN	ARMSTRONG	ALDRIN	BRAND
								HAISE
CSM/GNC	KAMMAN	WILLOUGHBY	WILLOUGHBY	KAMMAN	STRAHLE	COEN	KAMMAN	WILLOUGHBY
GUID&CTL	BORCHESS	KALISHEK	CANIN		COEN	BENSON	HUTCHINSON	
CSM EECOM	BURTON	LOE	AARON	BURTON	LOE	AARON	LIEBERGOTT	BURTON
	DUMIS				DUMIS		LOE	DUMIS
INCO						BROWN	HANCHETT	DEATKINE
FIDO	GREENE	GUTHRIE	SHAFFER	GREENE	DAVIS	PAVELKA	GREENE	SHAFFER
	BOSTICK			BOONE	BOSTICK	GUTHRIE		
RETRO	GRAVETT	PAYNE	LLEWELLYN	WEICHEL	PAYNE	BOSTICK	DEITERICH	LLEWELLYN
	LLEWELLYN			MASSARO				
GUIDO	HUTCHINSON	PAULES	PRESLEY	BALES	PAULES	PARKER	RUSSELL	PAULES
	BALES		RENICK	PARKER		TEAGUE		
BSE-1	WOLF	AMOS	BRADY			VAN RENSSALAER-HAMNER		
BSE-2	ROBINSON	PARK	HOOPER			STRAUSBAUGH		
BSE-3	STRAUSBAUGH	PUSCHINSKY				SHOOK		
LM CONTROL								
LM TELCOM						LM SYS-HANNIGAN		
LM EMU								
SURGEON	ZIEGLESCHMID	BEERS	HAWKINS	ZIEGLESCHI	BEERS	HAWKINS	ZIEGLESCHMID	BEERS
	CAPRENTIER							
SIMSUP	SHELLEY		SHELLEY			KOOS		
SOM COORD	PAYNE		BLALOCK			PAYNE		
EXPERIMENTS								
NETWORK	RANDALL	EGAN	RANDALL	OJALEHTO	EGAN	OJALEHTO	RANDALL	EGAN
SCC	WILSON	LONERO	WILSON	GUY	MONKVIC	MEYER	WILSON	YOUNG
M&O								
PROCEDURE	SUTTON	DRAUGHON	MOLNAR	KEYSER	TEMPLE	ARMSTRONG	KEYSEL	LARSEN
						WEYER	NICHOLSON	
FLT ACT OFF			STOUGH	HOLLOWAY	ANDERSON	HOLLOWAY	GUILLORY	STOUGH
PAO	FRULAND	MCLEAISH	HANEY	RILEY	JAMES	HANEY	MCLEISH	WARD
PAO ASST	WORRELL		BIGGS	STORYY	POWELL	POWELL	JOHNSON	STOREY
PHOTO			PATNESKY			PATNESKY	BIRD	BRAY
ARTIST						SHORES		
SIMULATION								
SPAN	ALDRICH	GRIFFIN	MAYNARD	ARABIAN	COHEN	SIMPKINSON	MAYNARD	TOMBERLIN
FCD-CSM			ALDRICH	HUTCHINSO	LIEBERGOT	ALDRICH	STRAHLE	CANIN
FCD-LM						HANNIGAN	EDELIN	WHITMORE
MSE						ZARCARO	VONEHRENFR	WARD
BLDG 45						ARABIAN	JONES	J LOWE
REMOTE SITE REP								
CRO	EALICK							
RKV	KUNDEL							
GUAM	TEMPLE							

(continued)

APOLLO

POSITION	APOLLO 9 3/3/69 11:00AM GUMBROP-SPIDER MCDIVITT-SCOTT-SCHWEIKART LC39A CSM104/LM3			APOLLO 10 5/18/69 12:49 AM EDT CHARLIE BROWN-SNOOPY STAFFORD-YOUNG-CERNAN LC39B CSN 106/LM4		
MISSION DIR FOD	HAGE KRAFT	LEE	MCMULLEN	HAGE KRAFT	LEE	MCMULLEN
FLIGHT DIR	KRANZ	GRIFFIN	FRANK	LUNNEY GRIFFIN	WINDLER	FRANK
AFD	LEWIS HARLAN	FENDELL	DRAUGHON	EALICK TEMPLE	HARLAN	FENDELL
CAPCOMS	ROSSA CONRAD	EVANS GORDON	WORDEN BEAN	DUKE MCCANDLESS	ENGLE	LOUSMA
CSM/GNC	HUTCHINSON KAMMAN	CANIN STRAHLE	COEN	KAMMAN COEN	STRAHLE WILLOUGHBY	CANIN
CSM EECOM	DUMIS	LIEBERGOTT	LOE	BURTON DUMIS	AARON	LIEBERGOTT
INCO	WAFFORD	BROWN	DEATKINE	HANCHETT	DEATKINE	BROWN
FIDO	REED BOSTICK	BOONE PAVELKA	KENNEDY SHAFFER	SHAFFER GREENE	GUTHRIE KENNEDY	STOVAL BOSTICK
RETRO	LLEWELLYN DEITERICH	SPENCER ELLIOTT	I'ANSON	WEICHEL DEITERICH	I'ANSON	LLEWELLYN ELLIOTT
GUIDO	FENNER PAULES	RENICK WELLS	PRESLEY PARKER	RUSSELL PAULES-BALES	FENNER TEAGUE	RENICK PARKER
BSE-1	BRADY-HAMNER			FRANK VAN "R"	BRADY	HANKS
BSE-2	DIXON			HARRIS	DIXON	PALMER
BSE-3	TOWNSEND			PARK	HOOPER	SHOOK
LM CONTROL	LODEN CRAVEN	CARLTON	WEGNER	WEGENER	LODEN	THORSON
LM TELCOM	PETERS MERRITT	PUDDY	KNIGHT	MERRITT	PETERS	KNIGHT
LM EMU	JOKI SAULTZ	YOUNG WALTON		T. WHITE SAULTZ	W. BATES WALTON	
SURGEON	HAWKINS	ZIEGLESCHMID	BEERS	HAWKINS	ZIEGLESCHMID	BEERS
SIMSUP	GRIFFITH			HONEYCUTT		
SIM COORD	CLEGG(AF)			HEINZ(AF)		
EXPERIMENTS						
NETWORK	RANDALL MEYER	EGAN VICE	STARCHURSK GONZALES	OJALEHTO YOUNG	MONKVIC GONZALES	WILSON DECOSMO
M&O PROCEDURES	MOLNAR THOMPSON	COVINGTON	KEYSER	ARMSTRONG PENNINGTON	LARSEN	WEYER
FLT ACT OFF	HOLLOWAY	O'NEILL	GARDNER	ANDERSON	O'NEILL	LINDSAY
PAO PAO ASST	HANEY RILEY	JAMES	WHITE	RILEY DUFF	WHITE	WARD
PHOTO ARTIST SIMULATION	PATNESKY	BIRD	BRAY	PATNESKY	TURNER	BIRD
SPAN	SIMPKINSON-COHEN MAYNARD		TOMBERLIN	SIMPKINSON-COHEN/MAYNARD-MORRIS		TOMBERLIN
FCD-CSM	ALDRICH	AARON	WILLOUGHBY	HUTCHINSON	BLAIR	LOE
FCD-LM	HANNIGAN	EDELIN	KEESLER	PUDDY	CARLTON	HANNIGAN
MSE	KUBICKI	GLANCY	WARD	KUBICKI	ZARCARO	WARD
BLDG 45	ARABIAN	JONES	DODSON	ARABIAN	JONES	DODSON
B/U CREW				COOPER	MITCHELL	EISELE

(continued)

APOLLO

POSITION	APOLLO 11				APOLLO 12			
	7/16/69 9:32AM CDT/LAND 7/20/69 4:17PM EDT		5 COLUMBIA-EAGLE		11/14/1969 YANKEE CLIPPER-INTREPID			
	ARMSTRONG/ALDRIN/COLLINS				CONRAD-GORDON-BEAN			
MISSION DIR	HAGE	LEE	MCMULLEN		MCMULLAN	HOLCOMB	STOUT	
FOD	KRAFT				KRAFT			
FLIGHT DIR	CHARLESWORTH	KRANZ	LUNNEY	WINDLER	GRIFFIN	FRANK	CHARLESW(	LUNNEY
	GRIFFIN	TINDALL						
AFD	PLATT	LEWIS	DRAUGHON	HARLAN	TEMPLE	KEYSER	PLATT	LEWIS
	KEYSER	TEMPLE	NICHOLASON		COVINGTON	JOHNSON	LEEPER	
CAPCOMS	DUKE	EVANS	MCCANDLESS		CARR	GIBSON	WEITZ	LIND
	LOVELL	ANDERS	MATTINGLY		SCOTT	WORDEN	IRWIN	(WASH)
	HAISE-SCHMITT	LIND	GARRIOTT					
CSM/GNC	COEN	WILLOUGHBY	STRAHLE	KAMMAN	WILLOUGHB'	STRAHLE	COEN	CANIN
CSM EECOM	DUMIS	AARON	LIEBERGOTT	BURTON	AARON	DUMIS	LIEBERGOT	BURTON
INCO	DEATKINE	BROWN	HANCHETT		THOMPSON	FUCCI	FENDELL	WEYER
	THOMPSON	FENDELL	FUCCI					
FIDO	REED	BOSTICK	GREENE	PAVELKA	REED	STOVAL	BOONE	GREENE
			SHAFFER	BOONE	BOSTICK	SHAFFER	PAVELKA	
RETRO	LLEWELLYN	WEICHEL	DEITERICH	ſANSON	ELLIOTT	SPENCER	WEICHEL	DEITERICH
				SPENCER	LLEWELLYN			
GUIDO	PRESLEY	RUSSELL	FENNER	RENICK	PRESLEY	RUSSELL	FENNER	RENICK
	PAULES	PARKER	BALES	MILL	BALES	WELLS	TEAGUE	PARKER
	WELLS				PAULES			
BSE-1	HAMNER-BRADY	FRANK VAN R	TOWNSEND		VAN "R"-HAMNER			
BSE-2	DIXON	HARRIS	PALMER		HARRIS			
BSE-3	HOOPER	PARKER	SIGLER		PARK			
LM CONTROL	CARLTON	WEGNER	THORSEN	LODEN	WEGENER	THORSON	LODEN	STRIMPLE
				STRIMPLE		CRAVEN		CARLTON
LM TELCOM	PUDDY	MERRITT	KNIGHT		MERRITT	PUDDY	PETERS	KNIGHT
	WATROS					WATROS		
LM EMU	JOKI	PETERS						
SURGEON	HAWKINS	ZIEGLESCHMII	BEERS		HAWKINS	DOCTOR "Z"	HUMBERT	
SIMSUP	KOOS				HONEYCUTT			
SIM COORD	BLALOCK				BLALOCK			
EXPERIMENTS	SAULTZ	WALTON						
NETWORK	STARCHURSKI	OJALEHTO	RANDALL	YOUNG	RANDALL	YOUNG	DECOSMO	
	EGAN	WILSON	MEYER	DECOSMO	SHEEHAN	MEYER	CHAPMAN	
ALSEP N.C.	VICE	GONZALES	HORSTMAN	CARR	VICE	GONZALES	CARR	
PROCEDURES	COVINGTON	ARMSTRONG	PENNINGTON	MOLNAR	ARMSTRONG	LAZZARO	FANELLI	BLACK
	LEEPER	LAZZARO						
FLT ACT OFF	GARDNER	HOLLOWAY	GUILLORY	LINDSAY	STOUGH	GARDNER	LINDSAY	O'NEILL
					HOLLOWAY			
PAO	RILEY	WHITE	WARD		MCLEISH	WHITE	WARD	
PAO 2	JOHNSON	WORRELL	BIGGS		RILEY			
PHOTO	PATNESKY	TURNER	BIRD	BRAY	PATNESKY	TURNER	BIRD	
ARTIST	PETER HURD	LAMAR DODD	FRANKLIN MAHON					
SIMULATION								
SPAN	SIMPKINSON	KUBICKI	ZARCARO		SIMPKINSON	COHEN	MORRIS	
FCD-CSM	HUTCHINSONB	ALDRICH	LOE	CANIN	BLAIR	LOE	KAMMAN	HUTCHINSON
FCD-LM	HANNIGAN	CRAVEN	KESSLER		WHITMORE	HANNIGAN	KESSLER	
MSE	PEACOCK	GLANCY	WARD		BLACKMER	TEVELDAS	PEACOCK	WARD
BLDG 45	ARABIAN	JONES	DODSON-MALLEY		ARABIAN	JONES	MALLEY	DODSON
B/U CREW								
ALSEP ENGR	SHARPE	KUNDEL	BASSHAM		BRIZZOLARA	GAUTHIER	MOSER	
		LUNAR LANDING TEAM MEMBERS						

(continued)

APOLLO

POSITION	APOLLO 13 4/11-4/17/70 ODYSSEY-AQUARIUS LOVELL/HAISE/SWIGERT				APOLLO 14 1/31-2/9/71 KITTY HAWK & ANTARES SHEPARD/ROSSA/MITCHELL			
MISSION DIR FOD	LEE SJOBERG	STOUT	HOLCOMB		LEE SJOBERG	HOLCOMB		
FLIGHT DIR	WINDLER	GRIFFIN	KRANZ	LUNNEY	FRANK LUNNEY	WINDLER	GRIFFIN	
AFD	DRAUGHON NICOLSON	TEMPLE HARLAN	LEEPER LEWIS	KEYSER	NICOLSON HARLAN	COVINGTON KEYSER	TEMPLE WOLFER	
CAPCOMS	KERWIN YOUNG	BRAND MATTINGLY	LOUSMA		FULLERTON EVANS	MCCANDLESS	HAISE	
CSM/GNC	COEN	WILLOUGHBY	STRAHLE	KAMMAN	COEN WATSON	KAMMAN	CANIN DEATKINE	
CSM EECOM	AARON	LIEBERGOTT	DUMIS	BURTON	AARON mCLENDON	DUMIS STARESINICH	LIEBERGOT MOON	
INCO	HANCHETT	PENNINGTON	GLINES	SCOTT	GLINES BLACK	ARMSTRONG	WEYER	
FIDO	GREENE BOSTICK	REED SHAFFER	STOVAL PAVELKA	BOONE	REED	BOONE	STOVAL GREENE	
RETRO	ELLIOTT LLEWELLYN	DEITERICH	SPENCER	WEICHEL	DEITERICH	SPENCER	TANSON ELLIOTT	
GUIDO	BALES WELLS PAULES	RUSSELL PARKER	FENNER TEAGUE	RENICK PRESLEY	RUSSELL	PRESLEY FENNER	RENICK WELLS	
BSE-1	BRADY	VAN RENS..R	HAMNER		VAN RENS...R	BRADY	HAMNER	
BSE-2	HANKS	HARRIS			HANKS	HARRINGTON	PALMER	
BSE-3	PARK	TOWNSEND			TOWNSEND	HOOPER	DAY	
LM CONTROL	WEGENER	THORSON	STRIMPLE	LODEN	STRIMPLE	WEGENER	THORSON	
LM TELMU	KNIGHT	PETERS	HESELMEYE	MERRITT	MERRITT	HESELMEYER	KNIGHT	
SURGEON	HAWKINS	ZIEGLESCHMII	POOL		HAWKINS	ZIEGLESHCMID	BAIRD	
SIMSUP	HONEYCUTT				HONEYCUTT			
SIM COORD	BLALOCK				BLALOCK			
EXPERIMENTS	KUNDEL	SAULTZ BRABANDT	GRIFFITH SHARPE					
NETWORK	YOUNG CARR	DECOSMO GONZALES	VICE HORSTMAN		DECOSMO MEYER	VICE SOETARET	HORSTMAN STARCHURSKI	
ALSEP N.C.	SEETAERT	DELLOSSO	CHAPMAN		DELLOSSO	GONZALES	YOUNG	
PROCEDURES	LAZZARO	THOMPSON	FUCCI		LAZZARO	PENNINGTON	SCOTT	
FLT ACT OFF	GARDNER ONEILL	LINDSAY HOLLOWAY	STOUGH	PIPPERT	GARDNER HOLLOWAY	STOUGH ONEILL	PIPPERT	
PAO	MCLEISH	WHITE	WARD		MCLEAISH	WARD	WHITE	
PAO 2	RILEY				RILEY			
PHOTO	PATNESKY				PATNESKY	TURNER	BIRD	
ARTIST								
SIMULATION								
SPAN	BROOKS	ROACH			ROACH	BROOKS		
FCD-CSM	ALDRICH	LOE	HUTCHINSON		ALDRICH	LOE	HUTCHINSON	
FCD-LM	HANNIGAN	PUDDY	WHITMORE		PUDDY	HANNIGAN	LODEN	
MSE	WARD	KOHRS	BLACKMER		STEWARD	BLACKMER	PEACOCK	
OPS MGR	KUBICKI	COHEN	SEVIER		KUBICKI	COHEN	SEVIER	
	MORRIS				NEBRIG	MORRIS	KOHRS	
B/U CREW								
ALSEP ENGR					KUNDEL	MOSER	MCDONALD	
RECOVERY	HOOVER	BULLOCK	BERTHIAUME		HOOVER	RICHMOND	SPONHOLZ	

(continued)

APOLLO

POSITION	APOLLO 15 7/26-8/7/71 ENDEAVOUR & FALCON SCOTT-WORDEN-IRWIN			APOLLO 16 4/16-27/72 CASPER & ORION YOUNG, MATTINGLY, DUKE		
MISSION DIR	PETRONE	LEE		PETRONE	LEE	WALLACE
FOD	SJOBERG			TINDALL		
FLIGHT DIR	GRIFFIN	WINDLER	LUNNEY	FRANK	KRANZ	GRIFFIN
			KRANZ	SHAFFER	PUDDY	HUTCHINSON/LEWI
AFD	TEMPLE			WOLFER	TEMPLE	
CAPCOMS	ALLEN	FULLERTON	HENIZE	PETERSON	FULLERTON	IRWIN
	MITCHELL	PARKER	SCHMITT	HAISE	ROSSA	MITCHELL
	SHEPPARD	GORDON	BRAND	HARTSFIELD	ENGLAND	OVERMEYER
CSM/GNC	COEN	KAMMAN	DEATKINE	COEN	CANIN	WATSON
	WATSON	CANIN		FITTS		LERDAL
CSM EECOM	AARON	LIEBERGOT	BURTON	AARON	DUMIS	MOON
	MCCLENDON	DUMIS	STARESINICH			
INCO	SCOTT	FENDELL	PENNINGTON	PENNINGTON	FENDELL	SCOTT
FIDO	GREENE	BOONE	STOVAL	STOVAL	GREENE	BOONE
	KENNEDY					
RETRO	DEITERICH	SPENCER	ELLIOTT	I'ANSON	DEITERICH	ELLIOTT
	PAVELKA	I'ANSON				
GUIDO	RUSSELL	RENICK	PARKER	RUSSELL	PRESLEY	RENICK
	PRESLEY					
BSE-1	VAN REENS...R	HAMNER	BRADY	VAN RENS...R		BRADY
BSE-2	HANKS	HARRINGTON	WARD	HANKS	PALMER	HARRINGTON
BSE-3	TOWNSEND	HOOPER	HOFFMAN	TOWNSEND	HOFFMAN	HOOPER
LM CONTROL	THORSON	WEGENER	STRIMPLE	STRIMPLE	THORSON	WEGENER
LM TELMU	HESELMEYER	KNIGHT	MERRITT	KNIGHT	PETERS	MERRITT
	PETERS			HESELMEYER		
SURGEON	HAWKINS	ZIEGLESCHMID	BAIRD		ZIEGLESCHMID	BAIRD
SIMSUP	GRIFFITH			KOOS		
SIM COORD	RIORDAN			HOLKAN		
EXPERIMENTS	GRIFFITH	SAULTZ	KOOS	GRIFFITH	KOOS	SAULTZ
OSO	DRAUGHON	BRIZZOLARO				
NETWORK	YOUNG	VICE	HORSTMAN	WRINKLE	YOUNG	HORSTMAN
	SHEEHAN	RANDALL	WRINKLE	MEYER	SHEEHAN	CONDITT
ALSEP N.C.	DELLOSSO	GONZALES	CARR	DELLOSSO	GONZALES	SOEATAERT
PROCEDURES	WEYER	GLINES	ARMSTRONG	WEYER	ARMSTRONG	GLINES
			RAMSELL		ANSON	RAMSELL
FLT ACT OFF	GARDNER		STOUGH	PIPPERT	GARDNER	STOUGH
		ONEILL	PIPPERT	ONEILL		HOLLOWAY
PAO	WHITE	WARD	RILEY	MCLEAISH	WARD	WHITE
PAO 2	COCKROFT	WORRELL	COZENS	RILEY	GREEN	
PHOTO	PATNESKY	TURNER/BIRD	BRAU	UNDERWOOD	PATNESKY	
ARTIST						
SIMULATION						
SPAN		ROACH		ROACH	BROOKS	HARLAN
FCD-CSM	ALDRICH	HUTCHINSON	STRAHLE/MOON	ALDRICH	HUTCHINSON	LIEBERGOTT
FCD-LM	LODEN	CARLTON	PUDDY	CARLTON	WHITMORE	LODEN
MSE	PEACOCK	BLACKMER	SEGNA	PEACOCK	BLACKMER	SEGNA
OPS MGR	KOHRS	KUBICKI	SEVIER	KOHRS	KUBICKI	SEVIER
				COHEN	MORRIS	MAYO
B/U CREW						
ALSEP ENGR						
RECOVERY	FILLEY	SNYDER	SPONHOLT	STULLKEN	BERTHIAUME	SPONHOLZ

(continued)

APOLLO

POSITION	APOLLO 17						
	12/7-19/72 AMERICA & CHALLENGER						
	CERNAN, EVANS, SCHMITT						
MISSION DIR	PETRONE	LEE	WALLACE				
FOD	TINDALL						
FLIGHT DIR	GRIFFIN	KRANZ	FRANK				
		HUTCHINSON	LEWIS				
AFD	TEMPLE						
CAPCOMS	FULLERTON	OVERMEYER	PARKER				
	ALLEN	SHEPARD	MATTINGLY				
	DUKE	ROSSA	YOUNG				
CSM/GNC	CANIN	COEN	WATSON				
CSM EECOM	DUMIS	AARON	MOON				
INCO	SCOTT	GLINES	ARMSTRONG				
	HANCHETT		FENDELL				
FIDO	GREENE	STOVAL	BOONE				
	EPPS						
RETRO	DIETERICH	TANSON	PAVELKA				
	VOLLINS						
GUIDO	PRESLEY	RUSSELL	MILL				
	FERGUSON	STONE	RENICK				
BSE-1	VAN RENSE...R						
BSE-2	HANKS	PALMER	HARRINGTON				
BSE-3	TOWNSEND	DEAN	HOOPER				
LM CONTROL	THORSON	STRIMPLE	LODEN				
LM TELMU	MERRITT	PETERS	KNIGHT				
SURGEON	ZIEGLESCHMID	POOL					
SIMSUP	HONEYCUTT						
SIM COORD	PAYNE						
EXPERIMENTS	HICKS	KOOS	GRIFFITH				
OSO	BRIZZOLARO	DRAUGHON	SAULTZ				
NETWORK	YOUNG	SHEEHAN	WRINKLE				
	CONDITT	MEYER	BAERD				
ALSEP N.C.							
PROCEDURES	RAMSELL	PENNINGTON	ANSON				
	WOOD	KENNEDY					
FLT ACT OFF	HOLLOWAY	PIPPERT	STOUGH				
PAO	RILEY	WARD	WHITE				
PAO 2	WORRELL	COZENS	WINSTON				
PHOTO	PATNESKY						
ARTIST							
SIMULATION							
SPAN	HONEYCUTT	BROOKS	ROACH				
FCD-CSM	LIEBERGOTT	LOE	KAMMAN				
FCD-LM	CARLTON	SHANNON	HANNIGAN				
MSE	BLACKMER	SEGNA	GLANCY				
OPS MGR	BATTEY	MORRIS	KUBICKI				
	SEVIER	KOHRS	LUNNEY				
B/U CREW							
ALSEP ENGR							
RECOVERY	SNYDER	CHAPUT	FILLEY				
	STULLKEN						

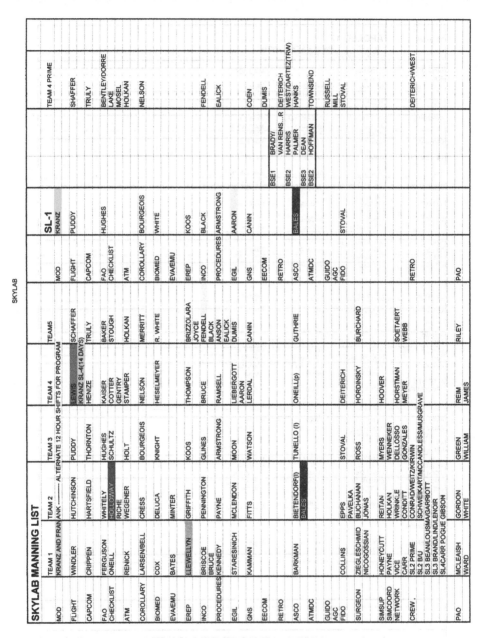

Figure A.2.5 Skylab Manning List.

Many thanks to former Flight Director Bill Reeves, President of the Manned Spaceflight Operations Association (MSOA) for compiling the names of people who served in the Mission Control Center.

Appendix 3
Women Flight Controllers and Mission Support

The history of women in flight operations is fairly recent. From the beginning of NACA, women primarily worked in administrative roles but there were many in the fields of science, mathematics, and computer support. From the beginning of NASA and the space program in 1958, flight operations was almost exclusively the domain of men with engineering and science degrees. They supported flight test pilots, who were also well educated.

During Mercury and Gemini, the mission control center was always staffed with men. However, when the new Houston Mission Control Center was built there were computer rooms and Staff Support Rooms (SSR) that required both administrative and computer support.

Some of the first women at MSC who supported missions in the early 1960s were Doris Folkes, Mary Shep Burton, Gloria B. Martinez, Cathy Osgood and Shirley Hunt, all of whom were in the Mission Planning and Analysis Division (MPAD). Women that supported flight operations organizations included Sue Erwin, Lois Ransdell, Connie R. Turner, and Maureen Bowen. They worked as secretaries for various members of the Mission and Flight Control teams.

By early Apollo, one would occasionally see several women in the SSRs but none in the MOCR unless they happened to enter with a message or some data. As missions became more complex and the software more sophisticated, you'd observe both contractors and NASA employees in the trajectory analysis areas such as the Flight Dynamics SSR, but only a few in the systems analysis areas such as the SPAN. Typically, the functions in the support areas were secretarial support, clerical support, math aides, data entry, and the traditional runners and messengers. Women with computer skills were supporting the RTCC, CCATS, and ACR.

NASA, and indeed the U.S. aerospace industry in general did not specifically seek to recruit women to work in specific traditional areas until the Equal Employment Opportunity Act of 1972. This prompted NASA to call in Nichelle Nichols of *Star Trek* fame as a recruitment consultant. She was the "face" of the recruitment drive. As a result, the hiring rates for women and minorities slowly increased.

While women began to hold more technical positions in the later years of the Apollo Program, none were considered to be "flight controllers" in the sense of having a position

© Springer International Publishing AG, part of Springer Nature 2018
M. von Ehrenfried, *Apollo Mission Control*, Springer Praxis Books,
https://doi.org/10.1007/978-3-319-76684-3

in the MOCR. The definition slowly changed to consider folks on the Manning List as being flight controllers.

But there was a hierarchy, as follows:

1. Those manning a console in the MOCR.
2. Those in the SSRs: e.g., Vehicle Systems, Spacecraft Analysis (SPAN), Flight Dynamics, Aeromed, Flight Director/Operations & Procedures and the Meteorological/Weather Room. Later there was a Science SSR for ALSEP and other Science Experiments on the lunar surface and in lunar orbit.
3. Functional Rooms: Recovery Operations Control Room, Simulation Control Room, RTCC, CCATS and ACR.

There were also other support areas, like the Mission Evaluation Room (MER) in Building 45, and the computer complex in Building 12. Although critical to the mission, the people that worked there were not considered to be flight controllers. Even so, there were only a couple of women out of more than a hundred working in the MER during Apollo.

Women with technical expertise would typically start in areas 2 and 3 above. Those with mathematical skills would gravitate toward the trajectory areas, and start in the Flight Dynamics Staff Support Room. Such was the case for Francis M. "Poppy" Northcutt who worked on Apollo trajectories. Those with computer skills would start in the RTCC, CCATS, or ACR.

By about the start of manned Apollo missions, more women were staffing the SSRs, RTCC, and CCATS; others represented contractors including PHO, GAC, TRW, ITT, Lockheed, and IBM, as well as NASA itself.

Some of the names on the Manning Lists include those below, but often only the first two initials of a name were listed, and looking back, it is difficult to tell whether some were women.

The following names, in alphabetical order, are for missions up to 1992 when MOCR 2 was abandoned. STS-53 was the last flight flown out of MOCR 2. The Apollo 7, Apollo-Soyuz, Skylab, and early Shuttle missions were controlled from MOCR 1. Note also that the number of women expanded once the Shuttle started flying in 1981. By the time the International Space Station was occupied in 2000, there were (and thankfully still are) women in many flight control positions, with some serving as Flight Director.

MOCR

- Karen M. Alig (FAO)
- Carolyn H. Blacknall (OIO)
- Michele Brekke (Payloads/Flight Director)
- Kathy V. Cannon (Payloads)
- Sharon B. Castle (Payloads)
- Lizabeth H. Cheshire
- Mary L. Cleave (CapCom-Astronaut)
- Carolynn L. Conley (FAO)
- Billie Deason (PAO)
- Bonnie Dunbar (GNC before becoming an astronaut)

- Marianne J. Dyson (FAO)
- Karen f. Ehlers (FAO)
- Anne F. Ellis (FAO)
- Diane L. Freeman (FAO)
- Linda M. Godwin (Payloads)
- Linda J. Hautzinger- Ham (Propulsion Officer-Flight Director 1991)
- Kathryn A. Havens
- Linda G. Horowitz (Aero)
- Jenny Howard (Booster)
- Angie Johnson (Payloads-1982)
- Cheevon B. Lau (FAO)
- Shannon W. Lucid (CapCom/Astronaut)
- Linda P. Patterson (GNC)
- Debbie T. Pawkett (Payloads)
- Barbara N. Pearson (EECOM)
- Sally Ride (Capcom before she was an astronaut)
- Janet K. Ross (PAO)
- Patraicia A. Santy (Surgeon)
- Barbara A. Schwartz (FAO)
- Ellen L. Schulman (Surgeon)
- Sharon R. Tilton (Surgeon)
- Gayle K. Weber (Guidance)

SSR
Technical including analysts, engineering and math aides:

- Jackie L. Barnes
- Susan M. Cardenas
- Judy Ferstl
- Holly Hensley
- Bonnie Hester
- Linda A. Holden
- Carolyn Huntoon (1972 Medical, Center Director from 1994–1996)
- Francis M. Northcutt (Apollo Flight Dynamics SSR)
- Janis Plesums
- Patricia A. Raby
- Sandra Smith

Administrative, including clerical, TTY operators, messengers, data:

- Jackie Allee
- Maureen E. Bowen
- Alandra Y. Brady
- Sharon Clarke
- Josephine C. Corey
- Donna R. Daughrity
- Patricia B. Dewey

- Marilyn H. Garzon
- Mary B. Gimsley
- Dorothy M. Hamilton
- Beverly Hobbs
- Joanne Hulo
- Phyllis Johnson
- Paula N. Jones
- Maxine Kitay
- Evelyn D. Langford
- B. Leschber
- C. Diann Merrell
- Olivia D. Merrell
- Ada W. Moon
- Frances P. Moore
- Christine M. Rizzo
- Edna Roberts
- S. Rogers
- Donna Sanders
- Dixie M. Scurlock
- Katherine Spencer
- Margery Weaver
- Dorothy M. Westover
- Diana R. Wiggins
- Maureen T. Ventura

RTCC

- Jackie L. Barnes
- Alva C. Hardy
- Holly Hensley
- Patricia A. Raby

CCATS

- Ann P. Crowder
- Dianne M. Ferguson
- Jackie E. Gregan
- Irene L. Hatfield
- Celeste E. Head
- Diannee L. Janecka
- Doris Kluge
- Claudeen Ledford
- Donna Tarpey

ACR

- Sandra DeWitt
- Diana Dutcher

- Mary Ann Lankford
- Penelope McGowen
- Mary McKee
- Patsy Ann Moore
- Delores Moorehead
- Faye Permenter
- Di Ann Sneed
- Connie R. Turner

Francis M. Northcutt	Marianne J. Dyson	Cheevon B. Lau
Carolynn L. Conley	Linda J. Ham	Michele Brekke

Figure A.3.1 Women flight controllers in the Mission Control Center.

Many thanks to Flight Directors Jeff Hanley, Linda Ham, and Milt Heflin for their inputs.

For further information see the References, including the book by Marianne J. Dyson, *A Passion for Space: Adventures of a Pioneering Female NASA Flight Controller*, 2016.

Appendix 4
Chronology of Events

The following is a chronological list of missions and control center events, as well as other events that place the spaceflight period into context. At times, we found it hard to appreciate everything that was going on!

One sometimes confusing point is the nomenclature of the Mission Operations Control Rooms (MOCR). The Mission Control Center-Houston in Building 30 of the Manned Spacecraft Center (since 1973 the Johnson Space Center) stands five stories tall but it has three floors. The first floor has the RTCC, CCATS and other miscellaneous display, communications and telephone equipment. MOCR 1 is on the second floor and MOCR 2 is on the third floor.

It is MOCR 2, along with several of the adjacent rooms that are being restored as a National Historic Landmark called the Apollo Mission Control Center.

10/4/57	Sputnik
10/1/58	NASA created from NACA
10/7/58	Space Task Group (STG) announced at Langley Research Center; formalized 11/1/58
12/18/58	SCORE launched; first voice satellite
4/1/59	Mercury astronauts selected; introduced to public 4/9/59
10/9/59	Chris Kraft introduces his Concept of Operations to the SETP
8/12/60	ECHO 1 launched
11/21/60	MR-1; first use of the Mercury Control Center
3/16/61	Goddard Space Flight Center dedicated
4/12/61	Yuri Gagarin; first man in space
5/5/61	MR-3; Alan Shepard becomes first American into space flying a ballistic arc
5/25/61	JFK challenges the nation to land a man on the Moon within the decade
7/21/61	MR-4; Gus Grissom
10/24/61	Manned Spacecraft Center (MSC) established in Houston, TX
1/20/62	John Glenn; MA-6 becomes the first American into orbit
3/1/62	Center Director Robert Gilruth moves from STG to MSC
4/19/62	IBM awarded the Gemini onboard computer contract
5/24/62	MA-7; Scott Carpenter
7/12/62	Telstar 1; first TV broadcast from space
7/30/62	STG transfer to MSC complete

© Springer International Publishing AG, part of Springer Nature 2018
M. von Ehrenfried, *Apollo Mission Control*, Springer Praxis Books,
https://doi.org/10.1007/978-3-319-76684-3

9/19/62	Overall design of the MCC
10/3/62	MA-8; Wally Schirra
10/15/62	IBM awarded the Ground Based Computing System, later named the Real Time Computer Complex (RTCC)
5/15/63	MA-9; Gordon Cooper makes the final Mercury flight
6/23/63	Philco contract for the MCC 2nd and 3rd floor MOCRs
7/18/63	MSC Phase I construction work complete
7/26/63	Syncom 2; first geosynchronous satellite
12/31/63	MSC Phase II construction work complete; start installation of equipment
4/8/64	Gemini 1; new MSFN for Gemini complete
8/19/64	Syncom 3; the world's first geostationary satellite; broadcast the Olympics from Tokyo
1/19/65	First passive use of MCC-H for Gemini 2
5/23/65	Gemini 3; Grissom and White; MOCR backup to Cape MCC
6/3/65	Gemini 4; White and McDivitt; MOCR 2 prime/operational
8/21/65	Gemini 5; Cooper and Conrad; MOCR 2
12/4/65	Gemini 7; Borman and Lovell; MOCR 2; rendezvous
12/15/65	Gemini 6-A; Stafford and Schirra; MOCR 2; rendezvous
2/26/66	Apollo Saturn 201; first unmanned Apollo test; MOCR 1
3/16/66	Gemini 8; Armstrong and Scott; first docking; emergency return; MOCR 2
6/3/66	Gemini 9-A; Cernan and Stafford; MOCR 2
7/5/66	Apollo Saturn 203; MOCR 1
7/18/66	Gemini 10; Young and Collins; MOCR 2
8/25/66	Apollo Saturn 202; MOCR 1
9/12/66	Gemini 11; Conrad and Gordon; MOCR 2
10/28/66	IBM Houston designs the RTCC
11/1/66	Gemini 12; Aldrin and Lovell; final Gemini; MOCR 2
1/27/67	Apollo 1 fire; witnessed by those in MOCR 1
11/9/67	Apollo 4; first unmanned Saturn V test; MOCR 2
1/22/68	Apollo 5; unmanned LM test; MOCR 1
4/4/68	Apollo 6; second unmanned Saturn V test; MOCR 2
10/11/68	Apollo 7; Saturn IB, first manned Apollo test; MOCR 1
12/21/68	Apollo 8; first manned Saturn V; circumlunar mission; MOCR 2
3/3/69	Apollo 9; MOCR 2
5/18/69	Apollo 10; MOCR 2
7/16/69	Apollo 11; first manned lunar launch; MOCR 2
7/20/69	First Lunar Landing
7/24/69	Return to Earth
11/14/69	Apollo 12; MOCR 2
4/11/70	Apollo 13; first deep space emergency; MOCR 2
1/31/71	Apollo 14; MOCR 2
7/26/71	Apollo 15; MOCR 2
4/16/72	Apollo 16; MOCR 2
12/6/72	Apollo 17; final Apollo lunar mission; MOCR 2
5/1/73	Skylab 1; Skylab launch; MOCR 1
5/2/73	Skylab 2; first crew-repair mission (28 days); MOCR 1
7/28/73	Skylab 3; endurance mission (59 days); MOCR 1
11/16/73	Skylab 4; endurance mission (84 days); MOCR 1
7/15/75	Apollo-Soyuz Test Project; MOCR 1
4/12/81	Space Shuttle; first manned flight; MOCR 1
11/12/81	STS-2; first reuse of a manned vehicle; MOCR 1
3/22/82	STS-3; R&D; only one to land at White Sands; MOCR 1

6/27/82	STS-4; last R&D; first DOD flight; MOCR 1
11/11/82	STS-5; first Shuttle controlled from MOCR 2
10/3/85	Apollo MOCR 2 designated National Historic Landmark
12/2/88	STS-27; first DOD Shuttle controlled from MOCR 2
12/2/92	STS-53 (DOD); last use of MOCR 2
7/20/17	Space Center Houston starts the Kickstarter Fund
8/21/17	Manned Spaceflight Operations Association established
1/25/18	Cosmosphere ships out the first batch of consoles for restoration and initiates refurbishment work in the MOCR
10/--/18	Cosmosphere returns the first batch of consoles and ships out the second set
7/16/19	50th anniversary of the Apollo 11 launch

Appendix 5
Photos and Graphics

"Deep space" EVA

Most people knew a little about the lunar surface experiments because they saw images of the astronauts deploying them or saw them off in the distance on their TV screens. Few people knew anything about the experiments going on in lunar orbit, or even knew there were a lot of scientists supporting the Mission Control Center. As reported in Section 5.3.5 two adjacent rooms supported the scientists and engineers monitoring those experiments. Section 7.5 lists all of the principal scientists.

While the CSM was in lunar orbit during the three final Apollo flights, there was a special bay in the side of the Service Module containing experiments that took photographs of the surface and gathered various other data. This was called the Scientific Instrument Module (SIM). In addition, a small package named the Particles and Fields subsatellite was released to remain in lunar orbit and report on conditions there. After leaving orbit to head home, one astronaut went out to retrieve the film and data cassettes from the SIM in an EVA conducted approx. 190,000 miles from Earth.

Apollo Lunar Surface Experiment Package (ALSEP)

One of my jobs at NASA was Chief of the Science Requirements and Operations Branch. In coordination with the scientists and Bendix engineers responsible for the experiments, we wrote documents that defined the scientific objectives of the lunar orbit and lunar surface experiments, and how the crews were to deploy and operate them.[1] I have found some great photos and descriptions of the Apollo 16 ALSEP experiments, which were designed to last 1 to 2 years but in some cases lasted much longer.

The ALSEP was unloaded from the LM descent stage storage compartment and transported approximately 95 meters southwest to a site designated Station 3/10. The ALSEP experiments were integrated into assemblages that were mounted on each end of a carrying bar. As the astronaut was making his way to the site, subpackage #2 fell off the bar. No damage was done though. There was a setback later when one of the astronauts snagged

---

[1] For a complete list go to: https://www.lpi.usra.edu/lunar/ALSEP/.

© Springer International Publishing AG, part of Springer Nature 2018
M. von Ehrenfried, *Apollo Mission Control*, Springer Praxis Books,
https://doi.org/10.1007/978-3-319-76684-3

Figure A.5.1 An artist's illustration of an astronaut retrieving the film canister from the SIM Bay of an Apollo spacecraft during a "deep space" EVA. Photo courtesy of NASA.

his boot in the cable to the heat flow experiment, ripping it off and thus rendering that experiment inoperable even before its emplacement was finished. The rest of the experiments were deployed without difficulty.

Emblems, Badges, Logos and Patches

Military squadrons have had emblems, badges, logos, and patches since at least WW-I. NASA has had a "logo" since 1958. Flight controllers have followed the tradition of making an emblem or patch for almost every program, mission, and organization. This tradition is not just used by manned spaceflight programs but by the JPL people too. Even corporations and institutions have them. The words are often used interchangeably, but I would think the word "emblem" represents the abstract meaning of an object such as a mission, organization, or squadron. I think that when you sew this to your jacket it becomes a badge or patch. After a while, various people use the words to suit themselves; for example a policeman carries a badge but his organization has a logo. A military man wears a badge as an insignia of his rank. Sometimes they are called ribbons, but they are emblems too. If you look up the word insignia in the dictionary it says that it is a badge or emblem. With that, here is the explanation of flight operations oriented emblems.

Let's start with a quote from page 376 of Gene Kranz's book, *Failure Is Not An Option*:

Bob McCall, in my belief, the premier artist of space, had been sitting on the step to the right of the flight director console, sketching during the final Apollo EVAs.

Figure A.5.2 A view of the Apollo 15 SIM Bay in lunar orbit. Photo courtesy of NASA.

He had designed the Apollo 17 crew patch. When Bob took a break for a cup of coffee, I joined him in the cafeteria. Like [Sig] Sjoberg, McCall's talent shone because of his sincerity and humility. As we talked, I don't think Bob was surprised when I asked him if he would design an emblem for the Mission Control team. I spoke emotionally, from my heart and gut, about the control teams and crews, and our life in Mission Control. "We fought and won the race in space and listened to the cries of the Apollo 1 crew. With great resolve and personal anger, we picked up the pieces, pounded them together, and went on the attack again. We were the ones in the trenches of space and with only the tools of leadership, trust, and teamwork, we contained the risks and made the conquest of space possible."

Figure A.5.3 This is the only image of Al Worden retrieving the film canister from the Apollo 15 SIM Bay. The 16-mm camera set up to record the operation jammed. Photo courtesy of NASA.

Figure A.5.4 A 16-mm camera view of Ken Mattingly retrieving the film canister from the Apollo 16 SIM Bay with the assistance of Charles Duke. Photo courtesy of NASA.

Figure A.5.5 Ron Evans retrieves the film canister from the Apollo 17 SIM Bay with the assistance of Jack Schmitt. Photo courtesy of NASA.

Over the next six months, McCall developed the emblem worn proudly by every subsequent generation of mission controller. He inscribed his final rendering of the emblem: "To Mission Control, with great respect and admiration, Bob McCall 1973."

In his NASA oral history, McCall reflected the credit back to Kranz. "He was very, very instrumental in creating the emblem. He was the one that asked me to do it, but also the one who really did most of the design. I just brought it together in a way that could be reproduced nicely."

The badge was changed many times. Firstly in 1983, to reflect the new Space Shuttle Program, then in 1989 to reflect the Challenger accident, then in 2003 to reflect the Columbia accident, and then again in 2012 and 2014.

Figure A.5.6 John Young starts deployment of the Apollo 16 ALSEP. In the foreground is the Lunar Surface Magnetometer (LSM) designed to measure the lunar magnetic field. The data could be used to investigate electrical properties of the subsurface and the interaction of solar plasma with the lunar surface. Photo courtesy of NASA.

At the top they all have the Latin inscription *Res Gesta Per Excellentiam*, for "Achieve through Excellence." The center symbolizes the responsibilities of the controllers, engineers and scientists in Mission Control who embrace each space program. The launch of a Shuttle represents the dynamic element of space. The four stars in the smoke of the boosters represent the four basic principles which guided the controllers during Mercury, Gemini, and Apollo; namely Discipline, Morality, Hardness, and Competence.

Five years later, following the tragic loss of Challenger and its crew of seven, the patch was revised, this time changing its central red banner to read "Mission Operations" so as to be inclusive of the entire team. The newly established Flight Operations Directorate (FOD) was the result of the merger of the flight crew and mission operations divisions. The directorate has responsibility for the astronauts' activities, as well as the planning and execution of their missions. It describes its mission as, "to select and protect our astronauts and to plan, train and fly human spaceflight and aviation missions." The Saturn V was removed and replaced with a comet as a memorial to the fallen astronauts.

It was the loss of Columbia in 2003 that inspired the next set of changes, thirty years after the patch was conceived. Graphic designer Michael Okuda updated the emblem to

Figure A.5.7 The Apollo 16 ALSEP deployed. The Passive Seismic Experiment (PSE) is in the foreground center, surrounded by its reflective skirt. The Central Station (C/S) is in center background with its antenna pointed towards Earth. It received commands from Earth, transmitted data, and distributed power to each experiment. The Radioisotope Thermoelectric Generator (RTG) is to the left. This was the power source for the ALSEP. The base of the RTG served as the base of the second ALSEP subpackage. One of the orange-topped anchor flags for the Active Seismic Experiment (ASE) is at the right. It enabled the internal structure of the Moon to be determined to a depth of several hundred feet, which in turn allowed scientists to infer the detailed structure of the upper kilometer of the lunar crust. After an astronaut had deployed three geophones by standing on them to drive them into the ground, he walked along the line, pausing periodically to place a "Thumper" on the ground and detonate a small explosive charge to propagate a shockwave to the geophones to investigate the immediate subsurface. Finally, a mortar package was deployed. After the LM had departed the Moon, this part of the experiment was to lob a set of larger explosive charges to varying distances out to 900 meters to probe deeper into the subsurface. Also notice the gold ribbon connectors to the Central Station. Photo courtesy of NASA.

Figure A.5.8 John Young deploys the Apollo 16 ALSEP. In the foreground is the Lunar Surface Magnetometer which measured how the strength of the Moon's magnetic field varied with time. Photo courtesy of NASA.

fix its star field at 17 blue stars: one for each of the fallen members of the Apollo 1, Challenger, and Columbia crews. He added the International Space Station to symbolize the permanent presence of man in space, and the white stars representing the principles of Mission Control were moved into the vector which extends from the rising Shuttle. A white star was placed on the Earth to represent Houston, the home of the mission operations team.

Okuda revisited his work in 2012, a year after the Shuttle program ended. The Shuttle and Station were removed from the emblem's central elements and added to the icons lining the "legacy ring" around the border. Taking their place were a stylized launch vector and an orbiting vehicle to represent the growing variety of American space vehicles in operation. Earth represents a view of America with a star to mark Houston. The 17 stars in the dark sky are a tribute to astronauts who are missing in flight. The comet pays homage

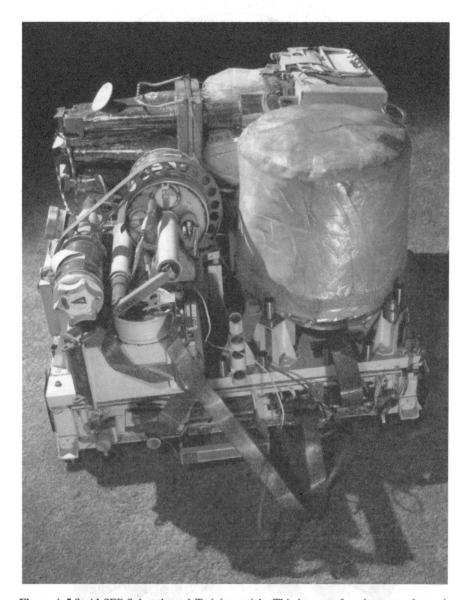

Figure A.5.9 ALSEP Subpackage 1 Training article. This is a set of equipment and experiments stowed together on an ALSEP pallet. The white cylindrical instrument is the Passive Seismic Experiment. To its left, is the Active Seismic Experiment. To the rear is the Lunar Surface Magnetometer. The experiments varied for each Apollo mission. This unit was used by Apollo 16 astronauts for Earth-based training of EVA procedures. It was transferred from NASA to the Smithsonian in 1975. It can be seen at the National Air and Space Museum in Washington, DC. Photo courtesy of the Smithsonian Institution.

Figure A.5.10 The First Patch, 1973.

Figure A.5.11 The 1983 Shuttle Patch.

Figure A.5.12 The first Flight Operations Directorate patch, 1988.

Figure A.5.13 The Post-Columbia patch, 2004.

Figure A.5.14   The Post-Shuttle patch, 2012.

to all those who have given their lives for the exploration of space. On the edge, symbols of the Mercury, Gemini, Apollo, Skylab, and ASTP programs.

There are slight differences in the 2014 patch compared to that of 2012. The launch vector plume, now split into three parts with an orbital path encircling it, represents astronauts at the forefront of space exploration. It also represents the dynamic elements of space, the initial escape from our environment, and the thrust to explore the universe. In the upper right of the emblem, the Moon and Mars represent our mission to lead the nation's permanent journey out of low Earth orbit. The Russian Space Station Mir was added to the legacy ring on the bottom border.

"We wanted to give people something that represents the U.S. space program and Johnson Space Center. We had Michael Okuda of Star Trek fame – but also well known for his NASA mission patches – come forward and design a special patch for us. It's absolutely beautiful," explained William Harris, CEO of Space Center Houston.

Okuda's commemorative design represents not only the physical hardware that played a key role in the Apollo program, but also the team of flight controllers. "I wanted to design something that was strongly evocative of the mission patch ethos, the mission patch being such a part of the NASA culture," said Okuda. "It came down to three elements – the big screens, the control consoles, and literally the touch of the flight controllers, who are the heart of Mission Control." He went on to observe, "I think the Apollo-era Mission Operations Control Room is one of the most important historical sites in 20th century

Figure A.5.15   The revised 2014 patch.

Figure A.5.16   The Apollo Mission Control Center Historical patch (left) and The Webster Challenge patch. Both were sponsored by Space Center Houston.

American history. It's the place where these amazing people faced astonishing challenges to literally send people where none had gone before. It is the symbol of what a great nation can do when we really try, when we commit to work together to challenge the impossible."

Figure A.5.17 The Manned Spaceflight Operations Association patch was also designed by Michael Okuda in 2017. It is based on the original McCall design themes using the Sigma and symbols of all the missions.

The Manned Spaceflight Operations Association (MSOA) was founded in the fall of 2017 to perpetuate the memory of those who came forward from all across America to fulfill President Kennedy's challenge, to send men to the Moon. This challenge was achieved in the Apollo Mission Control Center in Houston, Texas, which is now a National Historic Landmark.

The MSOA – which has applied for 501(c)(3) Corporation status – recognizes and honors those who planned, trained, and supported the many spaceflights that were controlled from this facility. This includes missions from Gemini 2 in 1964 through the entire Gemini program, the Apollo flights, Skylab, the Apollo-Soyuz Test Project and the early Space Shuttle flights through to 1992. These missions firmly established America's leadership in space in the sense that we are the only nation to have sent men to the Moon.

The MSOA supports the preservation and maintenance of the Apollo Mission Control Center facilities, including the consoles, displays and support equipment that were the primary focus of the operational personnel's working environments. This includes the preservation of support documents and other materials used at consoles during the missions. The Association intends to preserve their personal biographies, stories, photographs, and lessons for posterity via a website for that purpose.

The MSOA promotes the communication of the history of this historic facility and the benefits of spaceflight to society. It will seek to use this history to inspire new generations to educate and prepare themselves for future challenges in space. The MSOA will encourage knowledgeable speakers from this fellowship of men and women to inspire the next generation of scientists, engineers, technicians and administrators that will support future space programs. The association may seek to collaborate with other nonprofit organizations for educational, charitable, and scientific purposes.

Lunar Module Lift Off

One of the most remarkable and memorable images from the Apollo missions is the TV footage of the ascent stage of a Lunar Module lifting off from the Moon. How we obtained this is describe below.

NASA awarded contracts to build television cameras for Apollo to RCA and Westinghouse and both companies managed to build units for different missions that met NASA standards for weight, materials, and functionality. For the final three Apollo missions, RCA provided small, portable, color television cameras that could show the astronauts stepping off the LM onto the Moon, and then be installed on the Lunar Roving Vehicle to enable the viewers to accompany the astronauts on their exploration.

The cameras were very successful, capturing images of EVAs that included sample collection and the travails of moving and working in a space suit in one-sixth gravity. For the lunar liftoff, engineers had numerous calculations to make prior to the mission. The TV was on a pan and tilt unit that could be controlled directly from Earth via a large high-gain antenna on the rover. Since signals to and from Earth are delayed by a few seconds due to the 240,000 mile distance, mission engineers worked out a pre-programmed sequence designed to track an ascent stage as it lifted off from the Moon. Based on mathematical calculations, the LRV would be parked at a given distance and orientation from the LM. Then at just the right moment (allowing for the signal delay) INCO flight controller Ed Fendell in the MOCR would command the camera to start to tilt up at a rate that was calculated to follow the ascending craft.

On Apollo 15 the tilt mechanism malfunctioned and the camera didn't move, so the ascent stage moved out of the field of view within a few seconds. On the next attempt, the Apollo 16 astronauts parked the rover too close to the LM and this threw off the calculated tilt. We saw the vehicle lift off, but it soon vanished from sight. Apollo 17 was the charm. The rover was parked at the right distance from the LM. The tilt mechanism worked flawlessly, and we were able to watch the ascent stage until it was just a bright speck in the black lunar sky, on its way to rendezvous with the CSM. Even watching the video nowadays, half a century later, it is a magnificent sight.[2]

INCO Ed Fendell recalled in an oral history for NASA in 2000 how complex the procedure was:

> Now, the way that worked was this. Harley Weyer, who worked for me, sat down and figured what the trajectory would be and where the lunar rover would be each second

---

[2]You can watch Apollo 17 lift off from the Moon by going to YouTube and typing in "Apollo 17 LM Liftoff (inside and outside views)" as recorded from the TV camera on the rover and a 16-mm camera mounted in the window of the ascent stage.

Figure A.5.18 The LRV's final parking spot. The LM is in the distance. Photo courtesy of NASA.

Figure A.5.19 The Apollo 17 Lunar Module lifts off. Photo courtesy of the RCA camera on the rover, NASA and Ed Fendell.

Figure A.5.20   The ascent stage leaves the descent stage behind. Photo courtesy of the RCA camera on the rover, NASA and Ed Fendell.

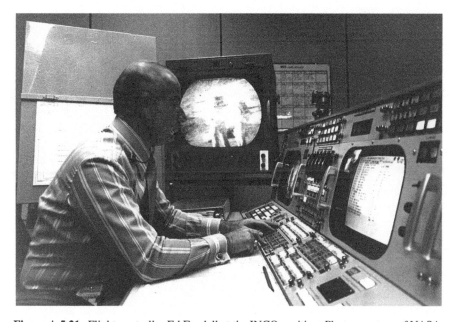

Figure A.5.21   Flight controller Ed Fendell at the INCO position. Photo courtesy of NASA.

as it moved out, and what your settings would go to. That picture you see was taken without looking at it at all. There was no watching it and doing anything with that picture. As the crew counted down, that's [an Apollo] 17 picture you see, as Cernan counted down and he knew he had to park in the right place because I was going to kill him, he didn't – and Gene and I are good friends, he'll tell you that – I actually sent the first command at liftoff minus three seconds. And each command was scripted, and all I was doing was looking at a clock, sending commands. I was not looking at the television. I really didn't see it until it was over with and played back. Those were just pre-set commands that were just punched out via time. That's the way it was followed.

Appendix 6
Quotes

Quotes from the Ancients
- "I have always felt it is my destiny to build a machine that would allow man to fly." – Leonardo da Vinci.
- "For once you have tasted flight you will walk the earth with your eyes turned skywards, for there you have been and there you will long to return." – Leonardo da Vinci.
- "There is nothing so removed from us to be beyond our reach or so hidden that we cannot discover." – Rene Descartes.

Quotes from Presidents
- "The whole world could see the awesome sight of the first launch of what is now the largest rocket ever flown. This launching symbolizes the power this nation is harnessing for the peaceful exploration of space." – President Lyndon B. Johnson said of the first flight of the Saturn V in 1967.
- "This is the greatest week in the history of the world since Creation." – President Richard M. Nixon to the Apollo 11 crew.
- "I have decided today that the United States should proceed at once with the development of an entirely new type of space transportation system designed to help transform the space frontier of the 1970s into familiar territory easily accessible for human endeavor in the 1980s and 1990s. This system will center on a space vehicle that can shuttle repeatedly from earth to orbit and back. It will revolutionize transportation into near space, by routinizing it. It will take the astronomical costs out of astronautics. In short, it will go a long way toward delivering the rich benefits of practical space utilization and the valuable spinoffs from space efforts into the daily lives of Americans and all people. Views of the earth from space have shown us how small and fragile our home planet truly is. We are learning the imperatives of universal brotherhood and global ecology – learning to think and act as guardians of one tiny blue and green island in the trackless oceans of the universe. This new program will give more people more access to the liberating perspectives of space. 'We must sail sometimes with the wind and sometimes against it,' said Oliver

© Springer International Publishing AG, part of Springer Nature 2018
M. von Ehrenfried, *Apollo Mission Control*, Springer Praxis Books,
https://doi.org/10.1007/978-3-319-76684-3

Wendell Holmes, 'but we must sail, and not drift, nor lie at anchor.' So with man's epic voyage into space-a voyage the United States of America has led and still shall lead." – President Richard M. Nixon, January 5, 1972.

Quotes from NASA Administrators

- "Spacecraft development, mission operations, and flight crew activities-in reviewing these areas of Apollo, I see one overriding consideration that stands out above all the others: Attention to detail. Painstaking attention to detail, coupled with a dedication to get the job done well, by all people, at all levels, on every element of Apollo led to the success of what must be one of the greatest engineering achievements of all time – man's first landing on the moon." – George Low on "What Made Apollo a Success."

- "Had the Lunar Orbit Rendezvous Mode not been chosen, Apollo would not have succeeded… without (John) Houbolt's persistence in calling this method to the attention of NASA's decision makers... [the agency] might not have chosen the Lunar Orbit Rendezvous." – George Low in 1982.

Quotes from NASA Center Directors

- "We try and plan for the unknowns. It's the unknown unknowns that you have concerns about. And that's what we are talking about, when you're talking about risk assessment." – Robert Gilruth during the Mercury Program.

- "I have learned to use the word "Impossible" with the greatest caution." – Wernher von Braun.

- "I have had the privilege of working with the giants of our profession, and I've had the good fortune to see future giants in the making. I'm confident our nation's future space endeavors will be in good hands." – Aaron Cohen.

Quote from Christopher Columbus Kraft, Jr.

- "….in planning the Apollo missions, much emphasis was placed on the demand for flexibility in the development program and responsiveness to changing needs. The dynamic conditions present in Apollo strongly influenced the mission planners in providing comprehensive alternate mission capability and flexibility in the ground and airborne flight software. Probably of more importance, however, was the capacity of the mission-planning team to react to major program readjustments, as evidenced typically by the Apollo 8 success. The effectiveness of this team, by using the process described here, is measured by the Apollo record." – Christopher C. Kraft, Jr. on "What Made Apollo a Success."

Quotes from Astronauts

- "If you eliminate the astronaut, you concede that man has no place in space." – Deke Slayton before the Society of Experimental Test Pilots October, 1959.

- "From the crew of Apollo 8, we close with good night, good luck, a Merry Christmas and God bless all of you, all of you on the good Earth." – Frank Borman.

- "The best way to face an unknown is to find out all you can about it in advance." – Gus Grissom.

- "If we die, we want people to accept it. We're in a risky business, and we hope that if anything happens to us it will not delay the program. The conquest of space is worth the risk of life." – Gus Grissom in an interview with Howard Benedict for the Associated Press in December 1966.

- "The more you train for something and the more you know about it, the more comfortable you are doing it." – Al Worden.
- "Houston, Tranquility Base here. The Eagle has landed." – Neil Armstrong, Commander Apollo 11.
- "That's one small step for a man, one giant leap for mankind." – Neil Armstrong, Commander Apollo 11.
- "It was a wondrous opportunity to be part of something historical. We just had a hard time comprehending what it would mean to other people, what it would mean to ourselves." – Buzz Aldrin.
- "The biggest benefit of Apollo was the inspiration it gave to a growing generation to get into science and aerospace." – Buzz Aldrin.
- "The important achievement of Apollo was demonstrating that humanity is not forever chained to this planet and our visions go rather further than that and our opportunities are unlimited." – Neil Armstrong.
- "When the history of our galaxy is written, and for all any of us know it may already have been, if Earth gets mentioned at all it won't be because its inhabitants visited their own moon. That first step, like a newborn's cry, would be automatically assumed. What would be worth recording is what kind of civilization we earthlings created and whether or not we ventured out to other parts of the galaxy." – Michael Collins.
- "OK Houston, we've had a problem!" – Jim Lovell.
- "...I'm on the surface; and, as I take man's last step from the surface, back home for some time to come – but we believe not too long into the future – I'd like to just [say] what I believe history will record. That America's challenge of today has forged man's destiny of tomorrow. And, as we leave the Moon at Taurus-Littrow, we leave as we came and, God willing, as we shall return, with peace and hope for all mankind. Godspeed the crew of Apollo 17." – Gene Cernan.

Quotes from Flight Directors

- "I think Kraft's name, Christopher Columbus, was entirely appropriate for this guy because he was the pioneer in Mission Control. He launched each one of the Mercury missions. But most important, he was the mentor, the teacher, the tutor for this first generation of young people who became known as Mission Controllers. He set the mold for everything that would be done thereafter; and in particular, he set the mold for the flight director and the flight director being able to take any action necessary for crew safety and mission success." – Gene Kranz about Christopher Kraft.
- "We had become very complacent about working in a pure oxygen environment. We all knew this was dangerous. Many of us who flew aircraft knew it was extremely dangerous, but we had sort of stopped learning. We had just really taken it for granted that this was the environment, and since we had flown the Mercury and Gemini program at this 100 percent oxygen environment, everything was okay. And it wasn't. And we had let the crew literally paper the inside of the spacecraft

with Velcro. I had each member of the control team write on the top of their black-boards in their offices "tough and competent" and that it could never be erased until we had gotten a man on the Moon. I believe that set the framework for our work in the weeks and months that followed." – Gene Kranz to the Flight Control Team after the Apollo 1 fire.

- "Achievement through Excellence" is the standard for the flight controllers' work. It represents an individual's commitment to a belief, to craftsmanship and persever-ance. (He goes on to say)- Sigma was chosen as the dominant element, representing the total mission team. In addition it represents the individual flight control teams from all programs past, present and future. Within the teams it represents all engi-neering, scientific and operations disciplines in support of the spacecraft. Sigma was chosen as the dominant element, representing the total mission team. In addi-tion it represents the individual flight control teams from all programs past, present and future. Within the teams it represents all engineering, scientific and operations disciplines in support of the spacecraft. The rocket launch represents the dynamic elements of space and the initial escape from our environment and the thrust to explore the Universe. The remaining elements are the Earth, planets and the stars. The Earth is our home and will forever be serviced by both manned and unmanned spacecraft in order to improve the quality of our present home. The stars and plan-ets represent a major source of study as well as the challenge of exploration for future mission control teams. The border contains symbols to represent the Mercury, Gemini and Apollo Programs, the three major programs that have been supported by the team. The four stars represent the present and future programs: Skylab, Apollo-Soyuz, Earth Resources and the Shuttle." – Gene Kranz describing the Mission Control emblem and patch.

- "To Mission Control, with great respect and admiration." – Bob McCall, 1973, after designing the Mission Control patch.

- "A functional organization has emerged which is flexible enough to meet unex-pected problems, but is structured enough to provide continuity of operation from mission to mission. The basic principles of flight control are not unique to manned space flight. They apply to any field where one can visualize malfunctions, docu-ment solutions, and rehearse the resulting actions. They could find use in any field where one monitors equipment or procedures by remote sensing devices. Application of the basic principles could increase efficiency in any field where one can write standard operating procedures. In the Apollo Program, they helped carry man to the lunar surface and bring him safely home again." – Gene Kranz and James Otis Covington on "What Made Apollo a Success."

- "If there should have been a lunar plaque left on the Moon from somebody in Mission Control or Flight Control – it should have been for Bill Tindall. Tindall was the guy who put all the pieces together, and all we did is execute them." – Gene Kranz.

- "From this day forward, Flight Control will be known by two words: Tough and Competent. Tough means we are forever accountable for what we do or what we fail to do. We will never again compromise our responsibilities... Competent means we will never take anything for granted... Mission Control will be perfect. When you leave this meeting today you will go to your office and the first thing you will do there is to write Tough and Competent on your blackboards. It will never be erased. Each day when you enter the room, these words will remind you of the price paid by Grissom, White, and Chaffee. These words are the price of admission to the ranks of Mission Control." – Gene Kranz in a speech to the Flight Control Branch in the aftermath of Apollo 1.

- "Apollo really did drive our industry. We were asking people to do things that were probably 10 or 20 years faster than they otherwise would have done. And they knew it. They stepped up to it and succeeded. Today's cell phones, wireless equipment, iPads and so on are a result of the fact that the country did this hi-tech thing and created this large portfolio of available technologies." – Glynn Lunney.

Quotes from Others

- "This is the goal: To make available for life every place where life is possible. To make inhabitable all worlds as yet uninhabitable and all life purposeful." – Dr. Hermann Oberth had written these words in his book *Man Into Space*, published in 1957.
- "The Apollo design philosophy has resulted in a highly reliable spacecraft capable of placing man on the moon and returning him to earth safely. Simple design practice, coupled with stringent technical and administrative discipline, has achieved this end." – Kenneth Kleinknecht on "What Made Apollo a Success."
- "The techniques used in Apollo for assessing systems performance reflect significant advancements over those used in previous manned programs. The method of handling flight anomalies, including the depth and extent of analysis, has been sufficient for the time and economy constraints imposed by the program. The Apollo concept has proved to be very effective in organizing many contractors and federal organizations into one central team for the real time support and post-flight evaluation of each mission. These concepts enabled the Apollo Program to advance at the rate required to achieve the national goal of landing man on the moon before 1970." – Donald D. Arabian on "What Made Apollo a Success."
- "There were several important spinoffs from this work. The (hundreds of) meetings were regularly attended by experts involved in all facets of trajectory control-systems, computer, and operations people, including the crew. Our discussions not only resulted in agreement among everyone as to how we planned to do the job and why, but also inevitably educated everyone as to precisely how the systems themselves work, down to the last detail. A characteristic of Apollo you could not help noting was just how great the lack of detailed and absolute comprehension are on a program of this magnitude. There is a basic communication problem for which I can offer no acceptable solution. To do our job, we needed a level of detailed understanding of the functioning of systems and software far greater than was generally available. Through our meetings, however, we forced this understanding. It was not easy, but we got it sorted out eventually-together." – Howard W. Tindall, Jr. on "What Made Apollo a Success."
- "At the start of a program, devise a thorough overall integrated test plan that includes all testing (including engineering and development, qualification, reliabil-

ity and life, pre-delivery environmental acceptance, pre-installation acceptance, installed system, altitude, prelaunch, and early unmanned flight tests). The plan should include as much testing as necessary to gain confidence in the hardware, the software, the test equipment, the test procedures, the launch procedures, and the flight crew procedures. The plan should provide for deleting unnecessary phases of testing as confidence grows." – Scott H. Simpkinson on "What Made Apollo a Success."

- "Voice, telemetry, command and tracking data acquired by the Goddard managed communications and tracking network represented some of the most critical information available to the flight controllers at their display consoles." – Lyn Dunseith, MPAD.
- "There is only one other event that may create as many advances in science and technology as the space program and that is war. I would rather go to space." – Charles A. Berry, MD., Flight Surgeon.

Poems for the Occasion

Voyage to the Moon
Wanderer in our skies,
dazzle of silver in our leaves and on our
waters silver, O
silver evasion in our farthest thought –
"the visiting moon," "the glimpses of the moon,"
and we have found her.
From the first of time,
before the first of time, before the
first men tasted time, we sought for her.
She was a wonder to us, unattainable,
a longing past the reach of longing,
a light beyond our lights, our lives – perhaps
a meaning to us – O, a meaning!
Now we have found her in her nest of night.
Three days and three nights we journeyed,
steered by farthest stars, climbed outward,
crossed the invisible tide-rip where the floating dust
falls one way or the other in the void between,
followed that other down, encountered
cold, faced death, unfathomable emptiness.
Now, the fourth day evening, we descend,
make fast, set foot at last upon her beaches,
stand in her silence, lift our heads and see
above her, wanderer in her sky,
a wonder to us past the reach of wonder,
a light beyond our lights, our lives, the rising
earth,

a meaning to us,
O, a meaning!
Archibald MacLeish for The New York Times, July 21, 1969

From Houston to Tranquility
(Upon viewing the abandoned Apollo Mission Control Center)
I do not see an empty room
I see people with missions to fly
And they see distant stars and moon
But now I just see neglect
Memories of dreams gone by.
And I don't want to hear sad excuses
Full of BS and half-truths
The last time I felt like this
Apollo Seventeen left behind a desolate Moon.
And now I stand in the shrine
And dream what was might be again
I dream what was might be again.
We planted the flag in the sands of Tranquility
Working as one, we saved lives from this place
We would restore it, this historic facility
And we want you to see, you to see this place.
by Spencer Gardner, Former Flight Activities Officer

Where is Our Purpose Now?
How well we remember
When we came in our youth
Naive yet realistic,
Dreamers but grounded.
We sought to participate in space
We sought to lead in space
For our country
For ourselves
For each other.
Where is our purpose now?
As for so many questions
We seek the answer in our past
The past yields no answer
But our past inspires us all – if we allow it.
This much we know
Our purpose will not be found in our dissension
It does not reside in pessimism
It resides not in the backwash of the age of space
It resides in our present
It awaits in our future.

Where is our purpose now?
Let us search for it together.
Let us fulfill it together.
by Spencer Gardner, Former Flight Activities Officer

The Room
We accepted the challenge given.
We worked hard; we were driven.
We were the privileged few.
We planned. We simulated.
By our mission we were stimulated.
We practiced both good and bad.
We learned from mistakes
Because that's what it took
If our nation's goal was to be had.
The crew – tip of the spear;
The mission – guided from here;
The room where we are,
The room with the view.
The world's point of contact
When lunar contact became fact.
We wanted to be in the room when it happened.
We want you to be in the room where it happened.
Open up the room where it happened.
Show everyone how history was made,
The use of the money they paid.
Man's dream came true
Man walked on the Moon
We watched from this room.
Now you too can feel
The dream made real.
2018 Spencer Gardner, Former Flight Activities Officer

Appendix 7
The Original MSC Contractors

"NASA...also owes debts to the contractors who imagined and implemented the high-tech systems that made the feats of manned spaceflight possible. What they created was an embodiment of a commitment to centralization coupled with high-technology that was particular to the Cold War era."

This was written by Layne Karafantis, author of *Under Control: Constructing the Nerve Centers of the Cold War* for her 2016 doctoral thesis at Johns Hopkins. She is now the Ames Research Center's historian.

How easy it is to forget who prepared the foundations of the Mission Control Center. Here is a brief history of those we should thank for their part in creating the "brick and mortar" of the now famous Building 30, as well as the equipment that it contained to enable NASA to reach the Moon.

In September 1961, the Fort Worth Division of the Army Corps of Engineers (ACOE) became the construction agency for the Manned Spacecraft Center in Houston, Texas. At the time, the future center was a cow pasture. Their first task was to hire an architecture/engineering (A/E) team to complete the initial design work. Twenty teams were considered for the contract and, after three rounds of reviews and rejections, an A/E team headed by Brown & Root, Inc., of Houston were selected. Partnered with them were the master planners Charles Luckman Associates of Los Angeles, California, and the architectural firms of Brooks & Barr of Austin, Texas; Harvin C. Moore, Houston, Texas; MacKie & Kamrath, Houston, Texas; and Wirtz, Calhoun, Tungate, & Jackson, Houston, Texas. The contract for almost $1.5 million was officially awarded in December 1961, and included general site development; master planning; design of the flight project facility, the engineering evaluation laboratory and the flight operations facility; and various site utilities.

Initial construction of the MSC was undertaken in three main phases. The contract for Phase I, preliminary site development, was awarded on March 29, 1962, to a joint venture of Morrison-Knudsen Construction Company of Boise, Idaho, and Paul Hardeman of Stanton, California. It was for $3,673,000. This work started in early April 1962, and was finished on

© Springer International Publishing AG, part of Springer Nature 2018
M. von Ehrenfried, *Apollo Mission Control*, Springer Praxis Books,
https://doi.org/10.1007/978-3-319-76684-3

July 18, 1963. The task included "overall site grading and drainage, utility installations including an electrical power system, a complete water supply and distribution system, sanitary and storm drainage systems, basic roads, security fence and street lighting."

Figure A.7.1 The building site for the Manned Spacecraft Center. Photo courtesy of NASA and Andrew "Pat" Patneski.

The invitations to bid for the Phase II contract of the construction, which was the first to include actual buildings, were distributed in early July 1962. By the time of contract award, the emphasis was placed on the Data Processing Center (Building 12). The contract award was in October 1962 with the joint venture of W.S. Bellows Construction Corporation, Peter Kiewit & Sons Corporation, and Ets-Hokin and Galvan, Inc., of Houston, Texas. The $4,145,044 contract was for Building 12, the sewage disposal plant, the central heating and cooling plant, the fire station, and a water treatment plant and associated building. The fire station was the first to be finished, in September 1963. The central heating and cooling plant was finished last in December 1963. During this same time period, Kaiser Engineers of Oakland, California, completed the design of the Mission Control Center.

The Phase III construction incorporated the largest grouping of buildings by one contract. The invitations to bid on this phase were issued on September 25, 1962, and listed ten buildings with an approximate total area of 760,000 square feet. As with Phase II, the statement of work was revised prior to the bid being submitted to include eleven office and laboratory buildings and the temperature and humidity control machinery for the entire

Figure A.7.2 The tall building in the center of the photo without windows is the new Mission Control Center in 1964. Photo courtesy of NASA.

site. Bidders were also invited to submit alternate proposals that incorporated additional facilities, which NASA was hoping to add to the contract if funding became available. On December 3, 1962, the contract went to the joint venture of C.H. Leavell and Company of El Paso, Texas, Morrison-Knudsen Construction Company, and Paul Hardeman. It was for roughly $19 million and specified eleven major facilities, including the project management building, the cafeteria, the flight operations and astronaut training facility, the crew systems laboratory, the technical services office and shop buildings, the systems evaluation laboratory, a spacecraft research lab and office building, and a data acquisition building. Funding for other facilities had become available by this time, so additional support buildings such as the shop building and warehouse were also listed. As per the contract, the buildings were to be ready for occupancy in 450 calendar days.

During the construction period, in October 1962, IBM was chosen to assemble the RTCC and Philco-Ford's Western Development Laboratories got the contract to provide all other electronics equipment such as the communications center, the flight simulator facilities, and the flight operations displays (the ones which are in the process of being restored).

In October 1963, the Logistics Division was the first to move into its facility, the Support Office (Building 419), and its shops and warehouse (Building 420). By the end of 1963, twelve further buildings were certified as operational. The major relocation to the new Center occurred between February and April 1964, and included moving into facilities such as the Auditorium and Public Affairs Facility (Building 1), the Flight Crew

Operations Office (Building 4), the Flight Crew Operations Laboratory (Building 7), the Systems Evaluation Laboratory (Building 13), and the Spacecraft Technical Laboratory (Building 16).

Dr. Robert Gilruth, the MSC Director, officially moved in on March 6, 1964, with his office in what was then Building 2 (it was later designated Building 1, the Project Management Building; the original Building 1, the Auditorium and Public Affairs Facility, immediately became Building 2). The Instrument and Electronics Laboratory (Building 15) was occupied in May. The occupation of Building 30, housing the MCC-H, was complete by at the end of June, when all leases on the temporary facilities expired. Gilruth declared an "Open House" for the weekend of June 6/7 and welcomed the public to view the new NASA Manned Spacecraft Center. Approximately 52,000 people toured the center and viewed displays that depicting the past, present, and future hardware of the space program.

Over a half century later, we salute the contractors and workmen that built the home of the "Apollo Mission Control Center," which is now a National Historic Landmark.

Appendix 8
In Memory of Our Colleagues

In addition to remembering the place where NASA spaceflights were controlled, we wish to remember all those who served in this now historic place and are no longer with us. Unfortunately, this list gets longer as time goes by. It used to be that we would be notified of the passing of our colleagues by Maureen Bowen, until she herself passed away. Now Bill Moon has taken up this unpleasant, yet much appreciated task. Thanks to you both.

Alford, Gay
Allday, Charles
Allen, James
Algranti, Joseph S.
Allen, Oliver
Arnold, John R.
Axford, Jon C.
Anderson, Bob
Anderson, Bill
Ankney, Walter Samuel
Appling, Jim
Armstrong, Lawrence L D.
Armstrong, Neil A.
Aubichon, Joyce E. (Gaddy)
Ballas, Bebe B.
Barker, Albert W.
Barnes, Phillip N.
Bastedo, William
Baxter, Hiram
Beatty, LaMarr D.
Becker, Robert W.
Beers, Sr., Kenneth, Dr.

Belew, Leland
Bell, Larry R.
Benner, R. L. (Bud)
Bennett, William J.
Benney, Jr., Alexie H. (Alex)
Benson, Jr., Richard B.
Berry, Joyce
Berry, Ron
Bilodeau, James
Blackburn, John
Blair, L. William
Bliss, George M.
Bohannon, Jackie W.
Bolender, Carroll H.
Bond, Alec M.
Booth, Lucille
Boudreau, Carole A.
Bourque, Donald J.
Bowen, John T.
Bowen, Maureen E.
Brady, James
Brandenburg, James R.

Brantley, Chester
Bray, Donald O.
Brzezinski, Michael S.
Brink, Jim
Britton, Bob
Brook, Dale
Brooks, Melvin F.
Broome, Douglas R.
Brown, Richard T.
Burton, Mary Shep
Burton, W. Clint
Burke, Roger A.
Byers, John P.
Canin, Lawrence S.
Capps, Charles R.
Carr, Earl V.
Cernan, Gene
Chaffee, Roger B.
Chanbellan, Rita
Chaput, Paul T.
Charlesworth, Clifford E.
Chesler, Mary

Chmielewski, Eugene B.

Coen, Gary E.

Cohen, Aaron C.

Colopy, Robert E., Cmdr.

Conditt, Julius

Conrad, Pete

Contois, George

Conway, George

Conwell, Jervy J.

Coons, M.D., D. Owen

Cooper, John H.

Cooper, Jr., L. Gordon

Coppens, Gerald L.

Corcoran, Jr., Lawrence O. (L.J.)

Coursen, J.

Cox, John

Craven, Jackson B.

Critzos, Chris C.

Culbertson, Ralph (Buddy)

Cutchen, Robert Eugene

Davis, Stuart L.

Dietlein, M.D., Lawrence F.

DiGenova, Frank

Dodson, Joe

Donbroski, Carline M.

Dudley, Nan

Dunseith, Lynwood C.

Durrett, Gene

Dye, Bobby B

Ealick, Perry

Easter, William B. (Bill)

Eisele, Donn

Eggleston, Jr., John Marshall

Ernull, Robert E.

Essl, William

Essmeier, Charles T.

Estes, Herbert S.

Evans, Ron E.

Faget, Maxime A.

Fato, Frank

Fenner, William E.

Ferguson, Gordon M.

Ferry, Johnny L.

Filley, Charles C. (Chuck)

Finley, Don

Finney, Dave

Fisher, Karla Garnuch

Fleming, Vincent A.

Flippin, Ann

Fowler, Richard

Frank, M. Pete

Franklin, Myles E.

Fucci, James

Fullerton, Gordon

Gallagher, Joseph H.

Garland, Benjamin J.

Gallagher, Joseph H.

Garland, Benjamin J.

Garman, John R. (Jack)

Garvin, Bill

Gaventa, Lawrence (Larry)

Germany, Daniel

Gilruth, Dr. Robert R.

Glenn, John H.

Glines, Alan C.

Gotsch, Wayne E.

Gordon, Richard F.

Graves III, Claude A.

Gravett, Faye

Greene, Jay

Greene, Joanne

Griffin, Marion

Grissom, Virgil I. (Gus)

Guerrero, Joe J.

Hackney, L. E. (Lou)

Hage, George H.

Hahne, Robert L (Bob)

Haithcoat, Kenneth E.

Hall, J. Leroy

Hammack, Jerome B.

Hammersly, Vernon C.

Hamner, R. Scott

Hand, Arthur A. (Art)

Haney, Leo

Haney, Paul

Hanssen, Viet

Harpold, Jon C.

Harris, William J.

Hartsfield, Henry W. (Hank)

Hatcher, John

Haugen, Kenneth R.

Haughton, John B.

Hawkins, M.D., Willard R.

Hazel, Frank John

Henry, Ellis W.

Henize, Karl

Henson, Orval E.

Hile, Bobby

Hoke, Frederick E.

Hoover, John E.

Hoover, Sr., Richard A.

Hopkins, Carroll E.

Howser, Lynn

Hrab, Walter

Huffstetler, Jr., William (Bill)

Hunter, Dan

Huston, Maj. Gen. Vincent G

Huss, Carl R.

Huss, Shirley (Yeater)

Hutson, Don E.

Hyle, Charles Tomas (Tom)

I'Anson, James E.

Irwin, James

James, Bennett W.

James, Joseph

Janes, Frank

Jenkins, Hank

Jenkins, Mark

Jenkins, Morris

Jenness, Martin D. (Marty)

Jezewski, Donald J.

Joerns, Jack Chase

Jones, Maj. Gen. David M.

Jones, Enoch

Jones, Sid

Johnston, Richard S.

Kelly, Thomas J.

Kennedy, Tom

Kuehnel, Helmut A.

Keune, Fred

Kimball, Garner R.

Kirbie, Richard

Kitay, Maxine

Klapach, Peter

Kleinknecht, Kenneth S.
Klingbeil, Sr., Dale L.
Koons, Fred
Kraak, Karen S.
LaCombe, Don
Lacy, William R.
Lamey, Jr. William C.
LaPinta, Charles K., MD
Larsen, Jr., Axel M. (Skip)
Laski, George
Lawson, Lee
Lee, Chester M. (Chet)
Leecraft, Bert
Legler, Robert
Lenoir, William B.
Lewis, Charles O.
Lindsey, Otho
Lineberry, Edgar C.
Linney, Rhea Q.
Lizza, Arthur E.
Llewellyn, John S.
Lockard, Sr., Daniel
Loftus, Joseph
Lopez, Sarah W.
LoPresti, Roy
Loree, M. Ray
Lovejoy, John I.
Low, George
Lowe, Merril A.
Lowery, Jerry L.
Loyd, Arnold J.
Lunde, Alfred N.
Lutes, Rex L.
Lynch, Arthur
Machell, Reginald M.
Maloney, Scotty
Martin, James A.
Marzano, Edwin F.
Mayer, Boyce
Mayer, John P.
Mayfield, Sam
Maynard, Owen
McBride, James
McCall, Robert T.
McCandless, II Bruce

McCown, Weldon (Gus) B.
McDonald, Donald J.
McDonald, Doyle J. (DJ)
McDonald, Kenneth D.
McElmurry, Thomas U.
McGathy, Jim
McLeaish, John
McWhorter, Larry B.
Meckley, Richard
Meyer, Grady
Michaud, Valerie
Milhoan, Jerry
Miller, Jim
Mitchell, Ed
Mitros, Edward F.
Molnar, Jr., William
Moore, Dale E.
Moore, James H.
Moore, Jr., Richard M.
Morris, Owen G.
Mosel, Duane K.
Moseley, Edward C.
Moser, James F.
Muehlberger, Dr. William
  R. (Bill)
Mueller, George
Murray, Bill
Muse, Gene
Myers, Robert L.
Neal, R. Terry
Nelson, Clair D.
Nelson, James F.
Nering, Paul D.
Neubauer, Milton (Jack) J.
Newman, Samuel R.
Nichols, Daniel
Noah, Newton
Noah, Wanda
Nolley, Joe W.
North, Warren J.
O'Briant, T. E.
O'Donnell, Robert A.
O'Neill, John W.
Overhouse, Raymond
Overmyer, Robert

Pace, (Chuck) Charles
Page, Floyd E.
Page, Thorton
Paine, Thomas O.
Patelski, K. J.
Patin, Quarance A (Quay)
Patnesky, Andrew R. (Pat)
Paules, Granville E.
Pavelka, Edward L.
Payne, Joe Dave
Pearson, Lee
Pennington, Granvil A.
Perner, Chris D.
Petrone, Rocco A.
Pettitt, George R.
Phillips, Samuel C.
Pike, Orlis V.
Pitts, Frank W.
Plesums, Janis
Poirier, Bob
Pool, M.D. Sam Lee
Present, Stu
Presley, Willard S.
Price, Thomas G.
Price, William E.
Prude, G. F.
Puddy, Donald R.
Quin, Edward E.
Quin, Matthew
Raines, Martin L
Rainey, Ed
Randolph, Ray
Ransdell, Lois
Ream, Howard (Bud)
Reid, Art
Reid, Doris W.
Reini, William A.
Ridge, Richard
Ritz, Bill
Roach, Jones W. (Joe)
Roberts, Harmon L.
Roberts, Tecwyn
Robertson, Richard (Robby)
Robinson, Willard D. (Robby)
Roosa, Stuart A.

Rose, Rodney G.
Rosenbluth, Marvin L.
Roy, Arda J.
Ruetz, Leroy L.
Sanborn, Sam
Sanchez, Abelino B. (A. B.)
Sanderson, Alan N.
Satterfield, J. M.
Saultz, Sr. James E.
Schiesser, Emil R.
Schirra, Walter P.
Schultz, Charles R. (Ray)
Schneider, William C.
Seaman, Charles K.
Segota, Peter
Sevier, John R.
Shaffer, Philip C.
Sharma, Herman S.
Sheaks, Larry E.
Shepard, Alan B.
Shoemaker, Dr. Eugene M.
Shook, G. R.
Simpkinson, Scott H.
Singer, Joe
Sjoberg, Sigurd A.
Skopinski, Ted H.
Slayton, Bobbie
Slayton, Deke
Smith, Ellis
Smith, Harry
Smith, John A.
Smith, Raymond L.
Snyder, Don G.
Spears, Glenn H.

Speier, William M. (Bill)
Spencer, Bobby T.
Spencer, Katherine J.
Steele, Willard
Stenfors, Hal
Stewart, Troy M.
Stone, Brock R. (Randy)
Stough, Charles L.
Straw, Hubert I.
Sturm, William E.
Stullken, Donald E.
Sulmeisters, Talivaldis K.
Swigert, Jack
Swim, Ron
Symons, Gilbert C.
Taylor, James J.
Taylor, Tommy
Taylor, William
Teague, Clyde
Theis, Richard A.
Thompson, Larry E.
Thorson, Richard A.
Tindall, Howard W.
Todd, William H.
Toleson, Robert
Tomberlin, James L.
Toups, L. Dean
Townsend, Don
Tunello, Rudy C.
Uljon, Linda
Vandervort, Quincy John
Vande Zande, Lyle
Vanos, Ted A.
Vice, Joe R.

Wadle, Richard C.
Wadleigh, Graydon (Grady)
Wafford, Larry
Walsh, Jack
Walton, Sr., Bruce H.
Warden, Henry E. (Hank)
Watkins, James
Webb, Donald E.
Webb, James E.
Weitz, Paul J.
Welch, Brian
Wheelwright, Charles D. (Chuck)
White, Jimmy C.
White, Lloyd H.
White, Lyle T
White, Robert C.
White, Robert T.
White, Ted A.
Whitmore, Charles V.
Wiseman, Donald G.
Williamson, F. G.
Willoughby, Briggs N. (Buck)
Wilson, Bill P.
Worley, Marian
Yardley, John F.
Yeakey, William (Bill)
Young, David A.
Young, John W.
Young, William C.
Zook, Herbert A.
Zwieg, Roger

Flight Crews Remembered

Although a flight controller is trained to respond to, and act upon, failures in his respective system or area of responsibility, in some cases there is nothing he can do. This terrible sense of helplessness occurred on all three of the space program tragedies. There was nothing that a flight controller could have done to avert the tragedies that overwhelmed the crews of Apollo 1 in 1967, STS-51L Challenger in 1986, or STS-107 Columbia in 2003. But the experience of witnessing one of these disasters leaves a mark and a vivid memory. I was the Guidance Officer on Apollo 1, and will never forget it. Some of my colleagues on duty at the time are in the list above. On January 27, 2017, NASA opened an "Apollo 1 Memorial" at KSC for the 50th anniversary of that first accident.[3]

After the accident, I drafted a letter to *Readers Digest* describing my feelings about the Apollo 1 tragedy. It was a cathartic exercise. Unfortunately JSC would not give me permission to send it. So, after half a century, here published for the first time, is what I thought about that day.

Mission Control's Darkest Hour
By Manfred von Ehrenfried
January 1967

My many years with the National Aeronautics and Space Administration have been filled with many memorable experiences. Most of them have been very pleasant; the kind of memories you chat about with friends over a mug of beer. Many of us in Flight Control have told and retold many a space story at our favorite drinking spots for post-mission parties – the Hofbraugarten and the Singing Wheel. But, many a beer fest will go by before a word is spoken about the last day of Apollo 1.

We had many simulations under our belts by that memorable day of January 27th. Plenty of "console" and "headset" time in previous Mercury and Gemini missions and in Apollo data flow tests, program checkouts, command tests, simulations, and much more. During those last several weeks before the accident there was a feeling of unreadiness in many a flight controller's mind. This was a new spacecraft, a new Apollo ground network, a new launch vehicle, new computer programs, and new positions within the control center for many of us. We still had a way to go before we would get that "flight ready" feeling we got in Gemini; the feeling you get when you are at home with your part of the system and with the job you have to do. We were confident that by launch day we all would be ready as we always have been in the past.

There were to be four Guidance Officers for Apollo 1. Two "Guidos" were to be on the console for launch phase and two for orbit phase. I was to come on in-orbit phase after spacecraft separation.

Each of us was assigned to one Flight Control team. We would work with that team during the simulations and tests as much as possible until we worked together efficiently; interpreting our data and passing critical bits of information to the right flight controller at the right time with the minimum amount of words and delay.

One of the Guidance Officers had come in early that day for the "Plugs Out" pad test in order to set up the Guidance Officers console and data displays which for this test was mostly command and telemetry. We had done a considerable amount of work with the command system in preparation for our part in the test. This consisted of formatting and transmitting several types of commands from Mission Control in Houston to the spacecraft at the Cape and then checking the spacecraft and ground's systems response via telemetry.

The Flight Dynamics Officers and Retrofire Officers were likewise preparing for their support in this test as were the Spacecraft Systems Engineers, the Network Controllers, the

---

[3]This can be seen by going to this link: https://www.nasa.gov/specials/dor2017/. You can also go to YouTube and type in "Tribute to Apollo 1".

Operations and Procedures Officers, the Booster Engineers, the Flight Surgeons, the Flight Directors and their assistants, and all of the Flight Control and Flight Support personnel.

One of the other "Guidos" and I relieved the first shift that day and started preparations for our support of the pad test. We checked the data that was prepared for us on the first shift and the notes on the test. Then we started checking our command procedures and equipment, our TV displays of telemetry, trajectory and general purpose data. We would always discuss what problems were still left open for us to chase down and what we needed to do in certain areas prior to the next simulation.

We were monitoring the various activities in the tests that concerned or interested us in particular and studying one problem or another during any holds, and making pertinent notes in our mission logs.

The communications seemed poor most of the day. The Blockhouse was having a little trouble "reading" the crew and a hold was called in order to remedy the problem.

I remember thinking to myself after listening to one transmission from the crew, "This communication is so bad their voices sound like they are changing frequency." But what I heard on the next transmission was clear enough, "Fire in the spacecraft!" I turned and yelled, "Did you hear that, something about fire in the spacecraft?" It seemed as though everyone was holding their earpiece into their ear so as not to miss a word. We couldn't see anything since no video was transmitted to Houston. Several other conversations were being heard on our communications panel which had nothing to do with the Blockhouse, so I punched off their monitors. My first reaction was not of concern so much as of sharp attention to the words that came into my ear. As the conversation on the Blockhouse loop became intense with the situation, a very sudden and chilling alarm came over me. I suddenly realized the extent of what was happening over a thousand miles away and that no one here could do a thing but listen to it too.

All of us hoped and prayed silently but as the minutes slipped by we knew it was-bad. Everyone in the Control Center was in a sort of shock after it was all over; Top management was called into the control room and advised of the situation or they were called on the phone. I watched these same people that I was used to seeing "all smiles and cigars" after most missions take the shocking news of what had happened to Apollo 1 and its crew. It seemed as quick as a bolt of lightning out of the blue but it did happen – it wasn't a dream or a simulation. Our Chief Flight Director was as stern and pale as I had ever seen him; quite a contrast to his appearance after a successful mission. The burden of his responsibility, even at hectic times, was always carried with an air of confidence which permeated the Control Center. This was a deep personal loss which showed itself as clearly as it was felt.

Our Division Chief and "Blue Team" Flight Director, was deeply moved. His premature grey hair seemed a predominant feature as I watched him during the discussions which followed. He was very quiet. Our prime Retrofire Officer was hard hit too. He and Gus Grissom had spent some great times together on previous Mercury and Gemini missions. "Retro" is an ex-marine who won the Purple Heart in Korea. He had some unforgettable experiences during the war. Tonight was one he'll never forget either.

I could tell that our prime Systems Engineer was feeling the loss deeply as he tried to maintain an air of calm as he replayed the telemetry data on his console, looking for clues to the accident.

My "constant console companion" didn't seem to move from his chair for what seemed like hours. This was to have been his first manned mission. He didn't believe what had happened so suddenly.

By 7:30 p.m. I couldn't stand to stay any longer. As I left the Control Center, I was in tears and was stopped by several people who had just heard the news and were rushing to the Control Center.

My wife, Jane, picked me up in front of work and by this time I couldn't hold back the tears, I could hardly breathe. As Jane drove rapidly off base, the Security Police stopped us for speeding. I jumped out of our car and walked back to him and told him that an accident had just happened and that I wanted to get home right away. He could sense that now wasn't the time for giving tickets and simply said "Go ahead."

Many thoughts went through my mind that night as I lay in bed trying to sleep. How could something that terrible happen and how could it happen that fast? I tried to recall from previous missions the many anxious moments in Mission Control. You can imagine the tense moments there must have been with the launch of America's first – Al Shepard. Or those alarming reports of hearing that Gus's Mercury spacecraft had sunk and he had to swim for it.

John Glenn's heat shield problem caused us some real concern too but we had a couple of hours to look at the problem and come up with a decision. We had control of the situation and although the reentry was "tense" to say the least, we did all the right things.

Scott Carpenter's retrofire, reentry and splash gave us some hectic moments but again everything came out without horrible results.

Wally Schirra's Sigma 7 environmental control system problem was the same sort of situation; one which was abnormal but under control; again, no problem.

We thought Gordon Cooper's Mercury spacecraft was suddenly going to "worms" near the end of the mission, but we came out of that situation with another success.

As all of the thoughts from the past missions went through my mind while I tried to sleep, I could not help but think about the guys on the launch pad and the job they did. Everybody in the Control Center heard the voice communications between the Pad Leader and the Test Conductor. We know these men and respect them for all they tried to do and the manner in which they conducted themselves. We hope that if and when our time comes in Mission Control, we can do as well; we are proud to work with such men.

I remember when I was in the White Room on Pad 19 looking at Grissom and Young's spacecraft "Molly Brown," the first Gemini spacecraft on top of our first Gemini launch vehicle – the Titan II. I got the horrible feeling then that there were so many things that could go wrong. Everything looked so strange, unfamiliar, and unknown that deep concern bordering on fear was my normal reaction. Why this "beast" didn't even have an escape tower on it. You had to use those horrible ejection seats. Even the propellants it burned were kind of eerie –they didn't look like flame and fire, you could see right through it. And when it staged; it fired-in-the-hole. This thing would never fly. But it did! They all did!

Then came the most fantastic moment in Mission Control; Ed White's EVA. That was really a heart thumper. We had worked for months – behind closed doors – writing flight plans, contingency mission rules – the works. That mission was all "pins and needles." Only the second Gemini flight and we were really outside the spacecraft!

And so it was throughout Gemini, some anxious moments and a lot of success. But Apollo 1 didn't even get off the ground! Mission Control had only played a pad support role. We didn't have time to analyze any abnormal situations. It was sudden and it was over! But it wasn't the end.

After the immediate realization of the accident comes the concern for the crew's families and for the Apollo Program. Both will be taken care of as best as can be. It feels good to say, after these many hard months, that something good will come from our great loss. I think spaceflight in general will become safer although risks may still be present in one insidious form or another. I think that everyone in Flight Control, in the Apollo Program Office, in Flight Crew Support, at the whole Center, and at each contractor facility feels more dedicated to the many difficult goals we have set for ourselves, for our country, for civilization. Spaceflight means more than a new adventure to us in Mission Control, it is more than going to work to make a living, and it is more than a way of life for us. It is a chance to contribute just a little bit to mankind, to be proud of ourselves and what we do from day to day. We have had "our darkest hour" and although we will have many pitfalls to overcome in the future we at least know that generations to come will be proud of what we have accomplished and the standards of achievement we have set for them.

The Ed Whites, Roger Chaffees, Gus Grissoms of Apollo 1, and all of the astronauts and cosmonauts who died before them and since them are true pioneers into another dimension of civilization, not only for their ultimate gift, their lives, but for their contributions while they were alive and possibly even more important, to those they inspire after their death. If the ultimate meaning of life is to have purpose; to make a contribution; to have the admiration and respect of your fellow man, to have the love of your family, and to enjoy life; then the Apollo 1 crew had it all!

Figure A.8.1 The AS-204 "Apollo 1" Crew. L-R: Senior Pilot Ed White, Command Pilot Gus Grissom, Pilot Roger Chaffee. Photo courtesy of NASA.

Figure A.8.2 STS-51L Challenger Crew. L-R: Commander Francis R. Scobee, Pilot Michael J. Smith, Mission Specialists Ronald McNair, Ellison Onizuka and Judith Resnik, Payload Specialist Gregory Jarvis, and Payload Specialist and Teacher Christa McAuliffe. Photo courtesy of NASA.

Figure A.8.3 STS-107 Columbia Crew. L-R: Mission Specialist David Brown, Commander Rick Husband, Mission Specialist Laurel Clark and Kalpana Chawla, Payload Commander Michael Anderson, Pilot William McCool, and Payload Specialist Ilan Ramon the first Israeli astronaut. Photo courtesy of NASA.

References

Books by Springer/Praxis authors

The Last of NASA's Original Pilots: Expanding the Space Frontier in the Late Sixties, by David J. Shayler and Colin Burgess, 2017

Friendship 7: The Epic Orbital Flight of John H. Glenn, Jr. by Colin Burgess, 2015

Liberty Bell 7: The Suborbital Mercury Flight of Virgil I. Grissom, by Colin Burgess, 2014

Freedom 7: The Historic Flight of Alan B. Shepard, Jr., by Colin Burgess, 2014

Moon Bound: Choosing and Preparing NASA's Lunar Astronauts, by Colin Burgess, 2013

How Apollo Flew to the Moon, by W. David Woods, 2011

Apollo 12 On the Ocean of Storms, by David M. Harland, 2011

Selecting the Mercury Seven: The Search for America's First Astronauts, by Colin Burgess, 2011

The First Soviet Cosmonaut Team: Their Lives, Legacies and Historical Impact by, Colin Burgess with Rex Hall, 2009

NASA's Moon Program, by David M. Harland, 2009

Paving the Way for Apollo 11, by David M. Harland, 2009

Robotic Exploration of the Solar System, by Paolo Ulivi and David M. Harland, 2009

Animals in Space: From Research Rockets to the Space Shuttle, by Colin Burgess with Chris Dubbs, 2007

The First Men on the Moon, by David M. Harland, 2007

Apollo: The Definitive Sourcebook, by Richard W. Orloff and David M. Harland, 2006

NASA's Scientist-Astronauts, by Colin Burgess with David J. Shayler, 2006

Space System Failures: Disasters and Rescues of Satellites, Rockets and Space Probe, by David M. Harland and Ralph D. Lorenz, 2005

Women in Space: Following Valentina, by David Shayler and Ian A. Moule, 2005

Water and the Search for Life on Mars, by David M. Harland, 2005

Apollo, by David J. Shayler, 2002

© Springer International Publishing AG, part of Springer Nature 2018
M. von Ehrenfried, *Apollo Mission Control*, Springer Praxis Books,
https://doi.org/10.1007/978-3-319-76684-3

Books by NASA Flight Controllers

The Kid From Golden: From the Cotton Fields of Mississippi to NASA Mission Control and Beyond, by Jerry Bostick, 2016

A Passion for Space: Adventures of a Pioneering Female NASA Flight Controller, Marianne J. Dyson, 2016

Go, Flight: The Unsung Heroes of Mission Control, 1965-1992 by Rick Houston and Milt Heflin, 2015

Highways Into Space, by Glynn Lunney, 2014

Below Tranquility Base: An Apollo 11 Memoir, by Richard Stachurski, 2013

From the Trench of Mission Control to the Craters of the Moon, by Glynn Lunney, David Reed and the Trench Flight Controllers, 2011

Failure is Not an Option: Mission Control From Mercury to Apollo 13 and Beyond, by Gene Kranz, 2009

Apollo EECOM: Journey of a Lifetime, by Sy Liebergot, 2003

Flight: My Life in Mission Control, by Christopher Kraft, 2001

Books by Others

Abandoned in Place: Preserving America's Space History, Roland Miller, 2016

Abandoned in Place: Interpreting the U.S. Material Culture of the Moon Race, Roger D. Launius, 2009

Apollo By The Numbers, Richard W. Orloff, 1996

Internet Links

https://www.hg.nasa.gov/alsj/mccfammanual.pdf (1965-67 Philco-Ford Familiarization Manual)

https://www.archives.gov/research/guide-fed-records/groups/255.html National Archives records of JSC (Can search on key words)

http://www.ibiblio.org/apollo/documents/tn-7685-apolloexperiencereport-flightcontrol needs.pdf (Can search on key words)

http://www.cr.nps.gov/history/online_books/butowsky4/index.html National Park Service Man in Space Theme Study-Dr. Harry A. Butowsky,1981

http://npshitory.com/publications/nhl/theme-studies/man-in-space/space.htm National Park Service History

https://www.lpi.usra.edu/lunar/missions/apollo/apollo_15/experiments/Lunar & Planetary Institute

https://www.lpi.usra.edu/lunar/documents/ Lunar & Planetary Institute

https://www.lpi.usra.edu/lunar/ALSEP/pdf/FC033.pdf ALSEP Flight Control Experiments Operations Plan-First Manned Lunar Mission, September 6, 1970

https://www.lpi.usra.edu/lunar/ALSEP/ For hundreds of reports on ALSEP

https://ntrs.nasa.gov/archive/nasa/casi.ntrs.nasa.gov/19790014808.pdf NASA Reference Publication 1036-ALSEP Termination Report, April 1979 https://ntrs.nasa.gov/archive/nasa/casi.ntrs.nasa.gov/19900011346.pdf Large Screen Display for the Mission Control Center, Ford Aerospace, Martin J. Skudlarek, 1989

https://www.nasa.gov/feature/apollo-1-crew-honored-in-new-tribute-exhibit New Apollo 1 "Ad Astra Per Aspera" exhibit for Grissom, White and Chaffee at Kennedy Space Center

https://www.nasa.gov/audience/forstudents/women-history-edu-index.html   NASA Women in Space

www.dutch-von-ehrenfried.com Link to "Dutch" von Ehrenfried's books

NASA Reports

Report of the Ad Hoc Working Group on Apollo Experiments and Training on the Scientific Aspects of the Apollo Program, 15 December 1963.

NASA TM X-58006 Apollo Lunar Landing Mission Symposium, June25–27, 1966.

NASA Goddard, The Manned Space Flight Network for Apollo, August, 1968.

NASA MSC Internal Note No. 69-FM-121, Apollo 10 Operations Review April 29, 1969.

Network Controller's Mission Report Apollo 11, MSC Flight Support Division August 15, 1969.

NASA MSC Internal Note No. 69-FM-121, Apollo 10 Operations Review April 29, 1969.

NASA SP-4307, "Suddenly Tomorrow Came...": A History of the Johnson Space Center 1957–1990, by Henry C. Dethloff, 1993.

NASA SP-287 What Made Apollo a Success? A series by the NASA History Office, Introduction by George Low, 1970.

NASA MSC Internal Note 70-FM-20, The Apollo 11 Adventure, February 5, 1970

Report of Apollo 13 Review Board, Appendix A Baseline Data: Apollo 13 Flight Systems and Operations, 1970

AS-508 MCC/MSFN Mission Configuration/System Description, March, 1970.

NASA MSC Flight Operations-Various Memorandum to Distribution for Apollo Manning Lists, 1964–1972.

NASA TN D-6855 Apollo Experience Report-Real-Time Auxiliary Computing Facility Development, by Charles E. Allday, June, 1972

NASA TN D-7993 Apollo Experience Report-Engineering and Analysis Mission Report by Robert W. Fricke, Jr., July 1975

NASA TM X-58131, Apollo Scientific Experiments Data Handbook, 1974

The Early Days of Simulation and Operations by Harold G. Miller, June 30, 2013

Contractor Reports

IMCC Systems and Performance Requirements Specification, Philco Western Development Laboratories WDL-TR-E120, September, 7, 1962

Facility Requirements and Criteria, Philco Western Development Laboratories WDL-TR-E112-3 September 7, 1962

PHO-FAM 001 Familiarization Manual: Mission Control Center Houston, June 30, 1967

Apollo Lunar Surface Experiments Package, ALSEP Familiarization Course Handout, NAS9-5829, Bendix Aerospace Systems Division, 1969

Historic Furnishings Report and Visitor Experience Plan: Apollo Mission Control Center National Historic Landmark, June 2015, National Park Service

PHO-TR155, MCC Operational Configuration Mission J1 Apollo 15, 1971

Mission Control Center/Building 30 Historical Documentation, by Archaeological Consultants, Inc. October, 2010

Glossary

ACHP	Advisory Council on Historic Preservation
ACOE	Army Corps of Engineers
ACR	Auxiliary Computer Room
AFD	Assistant Flight Director in the MOCR
AFRC	Armstrong Flight Research Center
AGS	Abort Guidance System
AIS	Apollo Instrumentation Ships
ALDS	Apollo Launch Data (Telemetry) System
ALSEP	Apollo Lunar Science Experiments Package
ALTDS	Approach and Landing Test Data System
AMR	Atlantic Missile Range
ANY	Antigua (tracking statin)
ARC	Ames Research Center
ARIA	Apollo Range Instrumentation Aircraft
AS	Apollo Saturn
ASIS	Abort Sensing and Implementation System
ASPO	Apollo Spacecraft Program Office
ASTP	Apollo-Soyuz Test Project
ATS	Atlantic Tracking Ship
ATV	Agena Target Vehicle
BDA	Bermuda (Tracking Station)
BJ	Big Joe
BMEWS	Ballistic Missile Early Warning System
BOOSTER	Call sign for the Booster Monitor in the MOCR
BSE	Booster Systems Engineer
CAL	Point Arguello, California (Tracking Station)
CAPCOM	Call sign for the Capsule Communicator
CCAFS	Cape Canaveral Air Force Station
CCATS	Communications, Command and Telemetry Support

© Springer International Publishing AG, part of Springer Nature 2018
M. von Ehrenfried, *Apollo Mission Control*, Springer Praxis Books,
https://doi.org/10.1007/978-3-319-76684-3

CDC	Control Data Corporation
CEO	Chief Executive Officer
CM	Command Module
COMTECH	Communications Technician
CNV	Cape Canaveral (tracking station)
COO	Chief Operating Officer
CRO	Carnarvon, Australia (Tracking Station)
CSD	Crew Systems Division
CSM	Command and Service Module
CSQ	Coastal Sentry Quebec (Tracking Ship)
CTN	Canton Island (Tracking Station)
CYI	Canary Island (Tracking Station)
DFRC	Dryden Flight Research Center (now AFRC)
DOD	Department of Defense
DPS	Descent Propulsion System (on the LM)
DSC	Dynamic Standby Computer
DSN	Deep Space Network
EASEP	Early Apollo Surface Experiment Package
ECS	Environmental Control System
EDS	Emergency Detection System
EECOM	Electrical, Environmental, Consumables in the MOCR
EGIL	Electrical, General Instrumentation, Life support in MOCR
EPS	Electrical Power System
ESSA	Environmental Science Services Administration
ETR	Eastern Test Range
EVA	Extra Vehicular Activity
FAA	Federal Aviation Administration
FCD	Flight Control Division
FCR	Flight Control Room
FD	Flight Director in the MOCR
FLIGHT	Flight Director's call sign in the MOCR
FIDO	Flight Dynamics Officer's call sign in the MOCR
FOD	Flight Operations Director(ate) also rep in MOCR
GAC	Grumman Aircraft Corporation
GBI	Grand Bahama Island (tracking station)
GLV	Gemini Launch Vehicle
GNC	Guidance Navigation Control System Engineer in MOCR
GNCS	Guidance Navigation Control System
GSFC	Goddard Space Flight Center
GT	Gemini Titan
GTI	Grand Turk (tracking station)
GUIDO	Guidance Officer in the MOCR (call sign GUIDANCE)
GYM	Guaymas, Mexico (Tracking Station)
HPC	Houston Petroleum Center
HPO	Historic Preservation Officer

IBM	International Business Machines
ICBM	Intercontinental Ballistic Missile
IGY	International Geophysical Year
IMCC	Integrated Mission Control Center
INCO	Instrumentation Communications Officer in the MOCR
IP	Impact Predictor (at KSC)
ITT	International Telephone and Telegraph
IU	Instrument Unit
JPL	Jet Propulsion Laboratory
JSC	Johnson Space Center
KSC	Kennedy Space Center
LC	Launch Complex
LJ	Little Joe
LED	Light Emitting Diode
LES	Launch Escape System
LH	Liquid Hydrogen
LM	Lunar Module
LOI	Lunar Orbit Insertion
LOR	Lunar Orbit Rendezvous
LOX	Liquid Oxygen
LRC	Langley Research Center
LRV	Lunar Rover Vehicle
LTA	Lunar Module Test Article
MA	Mercury Atlas
MAC	McDonnell Aircraft Company
MCC	Mercury Control Center, Mission Control Center
MCC-H	Mission Control Center-Houston (initially called IMCC)
MER	Mission Evaluation Room
MET	Mobile Equipment Transporter
MILA	Merritt Island Tracking and Data Station
MIP	Million Instructions Per second
MIT	Massachusetts Institute of Technology
MOA	Memorandum of Agreement
MOC	Mission Operations Computer
MOCR	Mission Operations Control Room
MOD	Mission Operations Director/Directorate
MOW	Mission Operations Wing
MPAD	Mission Planning and Analysis Division
MR	Mercury Redstone
MSC	Manned Spacecraft Center (now JSC)
MSFC	Marshall Space Flight Center
MSFN	Manned Space Flight Network
MSOA	Manned Spaceflight Operations Association
NACA	National Advisory Committee for Aeronautics
NASA	National Aeronautics and Space Administration

NASCOM	NASA Communications Network
NHL	National Historic Landmark
NPS	National Park Service
NR	North American Rockwell
NTHP	National Trust for Historic Preservation
O&P	Operations & Procedures Officer in the MOCR
OSO	Orbital Science Officer in the MOCR
OSW	Operations Support Wing
PAO	Public Affairs Officer/Commentator in the MOCR
PCM	Pulse Code Modulation
P&FS	Particles & Fields Subsatellite
PHO	Philco
PLHC	Public Lands History Center (Colorado State University)
PLSS	Portable Life Support System
PROCEDURES	Operations and Procedures Officer in the MOCR
RCA	Radio Corporation of America
RCS	Reaction Control System
RETRO	Retrofire Officer call sign in the MOCR
RFP	Request for Proposal
RSO	Range Safety Officer
RTACF	Real Time Auxiliary Computer Facility
RTC	Real Time Command Controller
RTCC	Real Time Computer Complex
S&AD	Science & Applications Directorate
S-IVB	Saturn number 4B (upper stage and Skylab workshop)
SCORE	Signal Communications by Orbital Relay
SCR	Strip Chart Recorder
SEB	Source Selection Board
SEP	Surface Electrical Properties
SHPO	State Historical Preservation Officer
SIM	Scientific Instrument Module
SLA	Spacecraft Lunar Module Adapter
SLS	Space Launch System
SLV	Saturn Launch Vehicle
SM	Service Module
SODS	Shuttle Operations Data Systems
SPAN	Spacecraft Analysis (SSR) in the MCC
SPS	Service Propulsion System
SSR	Staff Support Room
SSPDC	Space Shuttle Data Processing Complex
STADAN	Satellite Tracking and Data Acquisition Network
STG	Space Task Group
STS	Space Transportation System/Space Shuttle
SURGEON	Flight Surgeon in the MOCR
SYSTEMS	Call sign for the Systems Flight Controller

TDRSS	Tracking and Data Relay Satellite System
TEI	Trans Earth Injection
THC	Texas Historical Commission
TIC	Telemetry Instrumentation Controller
TLI	Trans Lunar Injection
TM	Telemetry
TRACK	Instrumentation Tracking Controller
TRW	Thompson Ramo Wooldridge
TTY	Teletype
UHCL	University of Houston at Clear Lake
UNIVAC	Universal Automatic Computer
UNIX	An AT&T computer operating system
USAF	United States Air Force
USB	Unified S-Band
USGS	United States Geodetic Survey
USN	United States Navy
USNS	USN Ship
UV	Ultra Violet
VIP	Very Important Person
VVR	Visitor Viewing Room
WDL	Western Development Laboratories (of Philco)
WFS	Wallops Flight Facility
WSMR	White Sands Missile Range

About the Author

Manfred "Dutch" von Ehrenfried II had the very good fortune to interview with the NASA Space Task Group the day before Alan Shepard was launched on the MR-3 mission. At the time, he had very little knowledge of Project Mercury and thought that because his degree was in physics he would be working in that area. As fate would have it, he was assigned to the Flight Control Operations Section under Gene Kranz, who became his supervisor and mentor. Most of his work for Project Mercury would be in the areas of mission rules, countdowns, operational procedures, and coordination with the remote tracking station flight controllers. During his first six months, he was in training to be a flight controller and spent MA-4 and MA-5 at the Goddard Space Flight Center learning communications between the Mercury Control Center and the Manned Space Flight Network.

His first mission as a flight controller was in the Mercury Control Center for John Glenn's MA-6 mission, learning the Procedures flight control position as understudy to Kranz. He then directly supported the Mercury orbital flights by Scott Carpenter, Wally Schirra, and Gordon Cooper.

Dutch supported the Gemini missions and was Assistant Flight Director for Gemini 4 to Gemini 7, which including the first EVA by Ed White and the first rendezvous in space by Gemini's 6 and 7. In 1966 he became a Guidance Officer on Apollo 1, and after the accident and stand down, became the Mission Staff Engineer on Apollo 7 and was backup in that position on Apollo 8. During this period, Dutch was also an Apollo Pressure Suit Test Subject. This afforded him the opportunity to test pressure suits in the vacuum chamber to the equivalent of 400,000 feet, including one test of Neil Armstrong's suit. He also experienced 9 g's in the centrifuge and flew aboard the zero-g aircraft. He had his own Apollo A7LB Skylab suit.

These experiences afforded him the opportunity to join the Earth Resources Aircraft Program. He became the first sensor equipment operator and Mission Manager on the high altitude RB-57F. As the sensor operator, he experienced working with scientists to operationally achieve their objectives. These flights required wearing a full pressure suit because they were generally at altitudes in the range 65–67,000 feet; one flight actually achieved 70,000 feet.

During 1970 and 1971 Dutch was the Chief of the Science Requirements and Operations Branch at NASA JSC. The Branch was responsible for the definition, coordination and

© Springer International Publishing AG, part of Springer Nature 2018
M. von Ehrenfried, *Apollo Mission Control*, Springer Praxis Books,
https://doi.org/10.1007/978-3-319-76684-3

Figure A.1 The author in late 1961 as a young STG Flight Controller. Center left: The author at the console to the left of Gene Kranz and George Low. Center right: The author testing Neil Armstrong's suit to an equivalent altitude of 400,000 feet in the vacuum chamber at the Manned Spacecraft Center. Bottom: The author wearing the A/P22S-6 full pressure suit required for the RB-57F. All Photos courtesy of NASA.

documentation of science experiments assigned to Apollo and Skylab. This included the Apollo Lunar Surface Experiment Packages (ALSEP) that were deployed on the Moon and experiments in lunar orbit and Earth orbit. The ALSEP packages included seismic sensors, magnetometers, spectrometers, ion detectors, heat flow probes, charged particle and cosmic ray detectors, local gravity instruments, and more. The lunar orbit experiments included the sensors and cameras carried in the Scientific Instrument Module (SIM) and the Particle and Fields subsatellites that were released in lunar orbit. The work also defined the procedures for the astronauts when deploying the packages and conducting experiments on the Moon and in lunar orbit.

Dutch also worked in the nuclear industry for seven years, and on the Space Station Program for ten years. In addition to writing several books reflecting on his experiences, he has been working in the finance and insurance fields for the past 20 years (www.dutch-von-ehrenfried.com).

Index

© Springer International Publishing AG, part of Springer Nature 2018
M. von Ehrenfried, *Apollo Mission Control*, Springer Praxis Books,
https://doi.org/10.1007/978-3-319-76684-3

Druck:
Canon Deutschland Business Services GmbH
im Auftrag der KNV-Gruppe
Ferdinand-Jühlke-Str. 7
99095 Erfurt